TOTAL CURVATURE IN RIEMANNIAN GEOMETRY

ELLIS HORWOOD SERIES IN MATHEMATICS AND ITS APPLICATIONS

Series Editor: Professor G. M. BELL, Chelsea College, University of London
(and within the same series)

Statistics and Operational Research
Editor: B. W. CONOLLY, Chelsea College, University of London

Baldock, G. R. & Bridgeman, T. — **MATHEMATICAL THEORY OF WAVE MOTION**
Beaumont, G. P. — **INTRODUCTORY APPLIED PROBABILITY**
Burghes, D. N. & Wood, A. D. — **MATHEMATICAL MODELS IN SOCIAL MANAGEMENT AND LIFE SCIENCES**
Burghes, D. N. — **MODERN INTRODUCTION TO CLASSICAL MECHANICS AND CONTROL**
Burghes, D. N. & Graham, A. — **CONTROL AND OPTIMAL CONTROL**
Burghes, D. N., Huntley, I. & McDonald, J. — **APPLYING MATHEMATICS**
Butkovskiy, A. G. — **GREEN'S FUNCTIONS AND TRANSFER FUNCTIONS HANDBOOK**
Butkovskiy, A. G. — **STRUCTURE OF DISTRIBUTED SYSTEMS**
Chorlton, F. — **TEXTBOOK OF DYNAMICS**
Chorlton, F. — **VECTOR AND TENSOR METHODS**
Conolly, B. — **TECHNIQUES IN OPERATIONAL RESEARCH**
Vol. 1: QUEUEING SYSTEMS
Vol. 2: MODELS, SEARCH, RANDOMIZATION
Dunning-Davies, J. — **MATHEMATICAL METHODS FOR MATHEMATICS, PHYSICAL SCIENCE AND ENGINEERING**
Eason, G., Coles, C. W., Gettinby, G. — **MATHEMATICS FOR THE BIOSCIENCES**
Exton, H. — **HANDBOOK OF HYPERGEOMETRIC INTEGRALS**
Exton, H. — **MULTIPLE HYPERGEOMETRIC FUNCTIONS**
Faux, I. D. & Pratt, M. J. — **COMPUTATIONAL GEOMETRY FOR DESIGN AND MANUFACTURE**
Goult, R. J. — **APPLIED LINEAR ALGEBRA**
Graham, A. — **KRONECKER PRODUCTS AND MATRIX CALCULUS: WITH APPLICATIONS**
Graham, A. — **MATRIX THEORY AND APPLICATIONS FOR ENGINEERS AND MATHEMATICIANS**
Griffel, D. H. — **APPLIED FUNCTIONAL ANALYSIS**
Hoskins, R. F. — **GENERALISED FUNCTIONS**
Hunter, S. C. — **MECHANICS OF CONTINUOUS MEDIA, 2nd (Revised) Edition**
Huntley, I. & Johnson, R. M. — **LINEAR AND NON-LINEAR DIFFERENTIAL EQUATIONS**
Jones, A. J. — **GAME THEORY**
Kemp, K. W. — **COMPUTATIONAL STATISTICS**
Kosinski, W. — **FIELD SINGULARITIES AND WAVE ANALYSIS IN CONTINUUM MECHANICS**
Marichev, O. I. — **INTEGRALS OF HIGHER TRANSCENDENTAL FUNCTIONS**
Meek, B. L. & Fairthorne, S. — **USING COMPUTERS**
Muller-Pfeiffer, E. — **SPECTRAL THEORY OF ORDINARY DIFFERENTIAL OPERATORS**
Nonweiler, T. R. F. — **COMPUTATIONAL MATHEMATICS: An Introduction to Numerical Analysis**
Oliviera-Pinto, F. — **SIMULATION CONCEPTS IN MATHEMATICAL MODELLING**
Oliviera-Pinto, F. & Conolly, B. W. — **APPLICABLE MATHEMATICS OF NON-PHYSICAL PHENOMENA**
Scorer, R. S. — **ENVIRONMENTAL AERODYNAMICS**
Smith, D. K. — **NETWORK OPTIMISATION PRACTICE: A Computational Guide**
Stoodley, K. D. C., Lewis, T. & Stainton, C. L. S. — **APPLIED STATISTICAL TECHNIQUES**
Sweet, M. V. — **ALGEBRA, GEOMETRY AND TRIGONOMETRY FOR SCIENCE STUDENTS**
Temperley, H. N. V. & Trevena, D. H. — **LIQUIDS AND THEIR PROPERTIES**
Temperley, H. N. V. — **GRAPH THEORY AND APPLICATIONS**
Twizell, E. H. — **COMPUTATIONAL METHODS FOR PARTIAL DIFFERENTIAL EQUATIONS IN BIOMEDICINE**
Whitehead, J. R. — **THE DESIGN AND ANALYSIS OF SEQUENTIAL CLINICAL TRIALS**

TOTAL CURVATURE IN RIEMANNIAN GEOMETRY

T. J. WILLMORE, B.Sc., M.Sc., Ph.D., D.Sc.
Professor of Pure Mathematics
University of Durham

ELLIS HORWOOD LIMITED
Publishers · Chichester

Halsted Press: a division of
JOHN WILEY & SONS
New York · Brisbane · Chichester · Toronto

First published in 1982 by

ELLIS HORWOOD LIMITED

Market Cross House, Cooper Street, Chichester, West Sussex, PO19 1EB, England

The publisher's colophon is reproduced from James Gillison's drawing of the ancient Market Cross, Chichester.

Distributors:

Australia, New Zealand, South-east Asia:
Jacaranda-Wiley Ltd., Jacaranda Press,
JOHN WILEY & SONS INC.,
G.P.O. Box 859, Brisbane, Queensland 40001, Australia

Canada:
JOHN WILEY & SONS CANADA LIMITED
22 Worcester Road, Rexdale, Ontario, Canada.

Europe, Africa:
JOHN WILEY & SONS LIMITED
Baffins Lane, Chichester, West Sussex, England.

North and South America and the rest of the world:
Halsted Press: a division of
JOHN WILEY & SONS
605 Third Avenue, New York, N.Y. 10016, U.S.A.

British Library Cataloguing in Publication Data
Willmore, Thomas J.
Total curvature in Riemannian geometry. –
(Ellis Horwood series in mathematics and its applications)
1. Geometry, Riemannian
I. Title
516'.93 QA685

Library of Congress Cataloguing in Publication No. 82-15670

ISBN 0-85312-267-9 (Ellis Horwood Limited – Library Edn.)
ISBN 0-85312-529-5 (Ellis Horwood Limited – Student Edn.)
ISBN 0-470-27354-2 (Halsted Press)

Printed in Great Britain by Unwin Brothers of Woking.

Table of Contents

Author's Preface

The purpose of this book is to introduce the reader to an interesting branch of geometry which has developed over the last thirty years, namely a study of invariants which arise from integrating various curvature measures over manifolds. In our treatment we have confined attention to smooth manifolds, though we are aware that the subject can be studied in the category of PL-manifolds and also in the topological category. Moreover, even in the smooth category we have paid particular attention to those curvature measures associated with the Lipschitz-Killing curvature and the mean curvature. This has implied omitting the integral representation of such important invariants as the Pontrjagin classes, the Hirzebruch genus, etc., but these lie outside the scope of an introductory book. Nevertheless by including a substantial bibliography, we hope that the reader will be encouraged to pursue his studies beyond the confines of the present volume.

I should like to thank my colleagues at the University of Durham for helpful discussions. I thank Mrs Joan Gibson for her excellent typing from my manuscript. I thank the staff of the publisher for their task in publishing from a mathematically complicated typescript. But, above all, I thank Ellis Horwood for a valued friendship over very many years, and in particular for his patience in awaiting the completion of the book.

Tom Willmore
University of Durham, May 1982

Chapter 1

Riemannian Geometry

1.1 INTRODUCTION

In this chapter we shall review the main properties of Riemannian geometry which we shall use throughout the book. Although we have tried to make our treatment reasonably self-contained, the reader will find it easier reading if he already possesses some familiarity with differential geometry, advanced calculus and some theory of integration in $\mathbb{R}^n$. The standard reference texts are Helgason (1978), and Kobayashi & Nomizu (1963) and (1969). However these standard texts are difficult for beginners (and others! ! !) and the reader may prefer the easier-to-read five-volume work by Spivak (1970), or the less ambitious book by the present author *An Introduction to Differential Geometry*, (1959).

1.2 DIFFERENTIABLE MANIFOLDS

Since we shall be primarily concerned with geometrical objects associated with differentiable manifolds, it is necessary to define precisely what we mean by this concept. Essentially, a differentiable manifold is a topological space equipped with a collection of charts. Each chart provides a one-to-one correspondence between open subsets of the manifold and open sets of the model space $\mathbb{R}^n$, and charts with overlapping domains provide consistency in differentiability properties.

More precisely, let M be a topological space. We assume that M satisfies the **Hausdorff separation axiom** which states that any two different points can be separated by disjoint open sets. [We make this assumption because ultimately we want to make M into a metric space, and this is only possible when M has the Hausdorff property]. We define an **open chart** on M to be a pair (U, ϕ) where U is an open subset of M and ϕ is a homeomorphism of U onto an open subset of $\mathbb{R}^n$.

We say that the Hausdorff space M has a **differentiable structure** of dimension m if there is given a collection of open charts (U_α, ϕ_α), where α belongs to some indexing set A, where $\phi_\alpha(U_\alpha)$ is an open subset of $\mathbb{R}^m$ such that the following conditions are satisfied:

M_1: $M = \bigcup_{\alpha \in A} U_\alpha$;

M_2: For each pair α, $\beta \in A$, the mapping $\phi_\beta \circ \phi_\alpha^{-1}$ is a differentiable mapping of $\phi_\alpha(U_\alpha \cap U_\beta)$ onto $\phi_\beta(U_\alpha \cap U_\beta)$;

M_3: The collection $(U_\alpha, \phi_\alpha)_{\alpha \in A}$ is a maximal family of open charts which satisfy conditions M_1 and M_2.

Remark 1 By **differentiable** in M_2 we mean infinitely differentiable, that is, the coordinates of a point in $\phi_\beta(U_\alpha \cap U_\beta)$ have continuous partial derivatives of arbitrary order as functions of the corresponding point in $\phi_\alpha(U_\alpha \cap U_\beta)$. We could, of course, consider restricting continuity of derivatives only for those of order k, but for our purpose this is not necessary. On the other hand, we could obtain an **analytic structure** by requiring the maps $\phi_\beta \circ \phi_\alpha^{-1}$ to be analytic.

Remark 2 In practice it is sufficient to check that conditions M_1 and M_2 are satisfied, because it is not difficult to prove that, when this is so, the family $(U_\alpha, \phi_\alpha)_{\alpha \in A}$ can be extended to a larger family in a unique way so that M_1, M_2 and M_3 are satisfied by this larger family.

We can now define a **differentiable manifold** of dimension m as a Hausdorff space with a differentiable structure of dimension m.

If M is a differentiable manifold, a chart (U_α, ϕ_α) where $\alpha \in A$ is often called a **local chart** on M or a **local coordinate system** on M. If $p \in U_\alpha$, and $\phi_\alpha(p) = (x_1(p), x_2(p), \ldots, x_m(p))$, then U_α is called a **coordinate neighbourhood** of p and the numbers $x_i(p)$ are called **local coordinates** of p.

An **analytic manifold** is defined in a similar way, "differentiable" in M_2 being replaced by "analytic".

A **complex manifold** of dimension m is defined by replacing $\mathbb{R}^m$ in the definition of a differentiable manifold by m-dimensional complex space C^m. Moreover, M_2 is modified by requiring that the m coordinates of $\phi_\beta \circ \phi_\alpha^{-1}$ be holomorphic functions of the coordinates of p, that is, each coordinate is expressible as a power series which converges absolutely in some neighbourhood of p.

Incidentally, we shall usually require our manifolds to be connected as topological spaces; and for manifolds connected implies arc-wise connected and conversely.

1.2.1 Differentiable functions on M

Let f be a real-valued function on M. We say that f is differentiable at p if there is a local chart (U_α, ϕ_α) with $p \in U_\alpha$ such that $f \circ \phi_\alpha^{-1}$ is a differentiable function on $\phi_\alpha(U_\alpha)$. Clearly this definition is independent of the particular chart chosen, because if (U_β, ϕ_β) is another chart with $p \in U$, then $f \circ \phi_\beta^{-1} = (f \circ \phi_\alpha^{-1}) \circ (\phi_\alpha \circ \phi_\beta^{-1})$.

We shall often use the notation $C^\infty(p)$ to denote the set of differentiable functions at p. If f is differentiable at each point $p \in M$ we say that f is **differentiable**. We denote the set of differentiable functions over M by $C^\infty(M)$. This set $C^\infty(M)$ forms an algebra over $\mathbb{R}$, the operations being

$$(\lambda f)(p) = \lambda f(p),$$
$$(f+g)(p) = f(p) + g(p),$$
$$(fg)(p) = f(p)\,g(p)$$

for $\lambda \in \mathbb{R}$; $p \in M$; $f, g \in C^\infty(M)$.

1.2.2 Open submanifolds

Let U be an open subset of the manifold M. We now show that there is induced over U a differentiable structure so that U becomes a differentiable manifold in its own right. Let M have a C^∞-structure given by the charts $(U_\alpha, \phi_\alpha)_{\alpha \in A}$. We consider the collection of open charts $(V_\alpha, \psi_\alpha)_{\alpha \in A}$ where $V_\alpha = U \cap U_\alpha$ and ψ_α is the restriction of ϕ_α to V_α. Clearly this makes U a differentiable manifold, which we call an **open submanifold**.

1.2.3 Extension of functions

Suppose we are given a differentiable function f over an open submanifold V of M. We ask whether this function can be extended to a differentiable function over the whole of M. We can always partly fulfil this requirement as is shown by the following theorem which we do not prove but refer the reader to Helgason (1978) page 7:

> **Theorem** 1. *Let V be an open submanifold of M, f a function in $C^\infty(V)$ and p a point in V. Then there exists a function $\tilde{f} \in C^\infty(M)$ and an open neighbourhood N of p with $N \subset V$ such that f and $\tilde{f}$ agree on N.*

Later on we shall use the existence of partitions of unity on manifolds for constructing global functions and structures out of ones defined locally. We could deal with this here but we prefer to introduce it later and make an illustrative application.

1.2.4 Product manifolds

Let M, N be differentiable manifolds of dimension m, n respectively. Let $(U_\alpha, \phi_\alpha)_{\alpha \in A}$, $(V_\beta, \psi_\beta)_{\beta \in B}$ be the corresponding collections of open charts for M and N. Let $\phi_\alpha \times \psi_\beta \colon (p, q) \longrightarrow (\phi_\alpha(p), \psi_\beta(q))$ denote the mapping $U_\alpha \times V_\beta \longrightarrow \mathbb{R}^{m+n}$, where $\alpha \in A$, $\beta \in B$. Then the collection $(U_\alpha \times V_\beta, \phi_\alpha \times \psi_\beta)_{\alpha \in A,\, \beta \in B}$ of open charts on the product $M \times N$ satisfies axioms M_1, M_2 and hence defines a manifold structure on $M \times N$, called the **product manifold**.

1.2.5 Exercises

(1) Show that $\mathbb{R}^n$ is a differentiable manifold.

(2) Let S^n denote the set of points in $\mathbb{R}^{n+1}$ given by

$$\sum_{i=1}^{n+1} x_i^2 = 1.$$

Let $N = (0, \ldots, 0, 1)$, $S = (0, \ldots, 0, -1)$, denoting respectively the north and south poles. Consider the two open charts $(S^n \setminus N, \pi_N)$,

$(S^n \setminus S, \pi_S)$ where π_N, π_S are stereographic projection from N and from S respectively, and show that these satisfy axioms M_1, M_2. Hence deduce that S^n is a differentiable manifold.

(3) Show that the set of all 2×2 non-singular real matrices is a differentiable manifold.
Hint: Use the fact that the determinant function is continuous to exhibit the set as an open subset of $\mathbb{R}^4$.

(4) Prove that the torus $S^1 \times S^1$ is a differentiable manifold.

1.3 VECTOR FIELDS

A **vector field** X on a differentiable manifold M is a derivation of the algebra $C^\infty(M)$. That is, X is a mapping $C^\infty(M) \longrightarrow C^\infty(M)$ such that

(i) $X(af+bg)=aXf+bXg$ for $a, b \in \mathbb{R}$; $f, g \in C^\infty(M)$;
(ii) $X(fg)=f(Xg)+(Xf)g$ for $f, g \in C^\infty(M)$.

The set of all vector fields over M, denoted by $\mathscr{D}^1(M)$, is a module over the ring $C^\infty(M)$. This follows because we define the vector field fX and $X+Y$ by

$$fX: g \longrightarrow f(Xg),\ g \in C^\infty(M),$$
$$X+Y: g \longrightarrow Xg+Yg,\ g \in C^\infty(M).$$

The reader should beware that fX is a vector field but Xf is a C^∞-function. If $X, Y \in \mathscr{D}^1(M)$, then it is easy to check that $XY-YX \in \mathscr{D}^1(M)$. We write $XY-YX=[X, Y]$, the so-called **Poisson bracket**. We shall also write $\theta(X)Y = [X, Y]$ and call the operator $\theta(X)$ the **Lie derivative** with respect to X.

Exercise

Verify that the bracket satisfies the Jacobi identity

$$[X, [Y, Z]]+[Y, [Z, X]]+[Z, [X, Y]]=0\ .$$

Alternatively prove that

$$\theta(X)([Y, Z])=[\theta(X)Y, Z]+[Y, \theta(X)Z]\ .$$

We state without proof the result, analogous to Theorem 1 for vector fields, and refer the reader to Helgason (1978) page 9,

Theorem 2. *Let Z be a vector field on an open submanifold $V \subset M$ and let $p \in V$. Then there exists a vector field $\tilde{Z}$ on M and an open neighbourhood $N \subset V$ with $p \in N$ such that $\tilde{Z}$ and Z induce the same vector field on N.*

We now prove the rather remarkable result that if X is a vector field defined on some coordinate neighbourhood U with local coordinates $x_1, x_2, \ldots, x_m$, then X is uniquely determined by its action on the coordinate functions. Let (U, ϕ) be a local chart on M, X a vector field on U and p some point in U. For $q \in U$ we write $\phi(q)=(x_1(q), x_2(q), \ldots, x_m(q))$. Let V be an open subset of U such that $\phi(V)$ is a star-shaped neighbourhood of

$\phi(p)=(a_1, a_2, \ldots, a_m)$. By star-shaped we mean that if $\phi(q)$ belongs to $\phi(V)$, then so do the points with coordinates $(a_i+t(x_i-a_i))$ for $0\leqslant t\leqslant 1$, $1\leqslant i\leqslant m$. If $f\in C^\infty(M)$ we write $f^*=f\circ\phi^{-1}$. Then we have as an identity

$$f^*(x_1, \ldots, x_m)=f^*(a_1, \ldots, a_m)+$$

$$\int_0^1 \frac{\partial}{\partial t} f^*(a_1+t(x_1-a_1), \ldots, a_m+t(x_m-a_m))dt$$

$$=f^*(a_1, \ldots, a_m)$$

$$+\sum_{j=1}^{m}(x_j-a_j)\int_0^1 f_j^*(a_1+t(x_1-a_1), \ldots, a_m+t(x_m-a_m))dt$$

Here we have used the standard rule foı differentiating a "function of a function", and we have used the notation f_j^* to denote the partial derivative of f^* with respect to the jth argument.

Writing the above relation with reference to points on M we get

$$f(q)=f(p)+\sum_{i=1}^{m}(x_i(q)-x_i(p))g_i(q), \quad q\in V$$

where $g_i\in C^\infty(V)$ for $1\leqslant i\leqslant m$ and

$$g_i(p)=\left(\frac{\partial f^*}{\partial x_i}\right)_{\phi(p)}.$$

However properties (i) and (ii) of a vector field give the result that a vector field maps a constant function into the zero function. Using this, we have

$$(Xf)(p)=\sum_{i=1}^{m}\left(\frac{\partial f^*}{\partial x_i}\right)_{\phi(p)}(Xx_i)(p) \text{ for } p\in U.$$

Now the mapping $f\longrightarrow\left(\frac{\partial f^*}{\partial x_i}\right)\circ\phi$ for $f\in C^\infty(U)$ is a vector field on U which we denote by $\frac{\partial}{\partial x_i}$. Then we have

$$X=\sum_{i=1}^{m}(Xx_i)\frac{\partial}{\partial x_i} \text{ on } U.$$

Thus we have justified our claim that X is determined by its action on the x_i. Moreover, we have proved that the m vector fields $\frac{\partial}{\partial x_i}$, $i=1, 2, \ldots, m$ form a basis for the module $\mathscr{D}^1(U)$.

1.3.1 Tangent vectors

We are now in a position to define what we mean by a tangent vector to M at p. Clearly we cannot use euclidean geometry to guide us, as the concept of the limit of the chord joining p to q is meaningless in our case. That is why we have had to use a more complicated and abstract approach. Nevertheless it will be seen that the tangent vectors defined in this way do have all the properties that one intuitively attributes to "tangent vectors".

Denote by $C^\infty(p)$ the space of real-valued functions which are C^∞ in some neighbourhood of p. Let X be a vector field over M and denote by X_p the linear transformation given by $X_p: f \longrightarrow (Xf)(p)$. The set $\{X_p: X \in \mathscr{D}^1(M)\}$ is called the **tangent space** to M at p, and its elements are called tangent vectors. We denote the tangent space to M at p by the symbol M_p. Clearly the tangent space to M at p is a vector space over $\mathbb{R}$. Moreover, the m vectors e_i, $i = 1, \ldots, m$ defined by

$$e_i: f \longrightarrow \left(\frac{\partial f^*}{\partial x_i}\right)_{\phi(p)}, \; f \in C^\infty(M) ,$$

are linearly independent and form a basis for the tangent space, which is therefore m-dimensional. If we write $X^i = X(x_i)$, then we have

$$X = \sum_{i=1}^{m} X^i e_i \; .$$

Clearly, our treatment of tangent vectors is analogous to that of directional derivative in elementary calculus. The basis (e_i) is called a **coordinate basis**.

1.3.2 Differential 1-forms

Let $\mathscr{D}_1(M)$ denote the dual of the module $\mathscr{D}^1(M)$ over $C^\infty(M)$. The elements of $\mathscr{D}_1(M)$ are called **differential 1-forms**.

We denote by (dx^i) the dual basis corresponding to (e_i). Then any differential 1-form ω can be written

$$\omega = \sum_{i=1}^{m} \mu_i dx^i$$

where the μ_i belong to $C^\infty(M)$.

The set of elements of $\mathscr{D}_1(M)$ restricted to p is called the space of **covariant vectors** at p, and is denoted by M^*_p.

1.4 TENSOR ALGEBRA

Let V be an m-dimensional vector space over $\mathbb{R}$ and let V^* denote its dual. Then a tensor T of type (r, s) is a multilinear mapping of $V^* \times V^* \times \ldots \times V^* \times V \times V \times \ldots \times V \to \mathbb{R}$. Note that there are r entries from V^* and s entries from V. It is convenient to regard $f \in \mathbb{R}$ as a tensor of type $(0, 0)$. Let

(e_i) be a basis of V and let (ω^i) be the dual basis. Write

$$T^{i_1 \ldots i_r}_{j_1 \ldots j_s} = T(\omega^{i_1}, \omega^{i_2}, \ldots, \omega^{i_r}, e_{j_1}, e_{j_2}, \ldots, e_{j_s}) .$$

Then the real numbers $T^{i_1 \ldots i_r}_{j_1 \ldots j_s}$ are called the components of T relative to the basis (e_i). The multilinearity ensures that T is completely determined by its action on the basis vectors (e_i), and the dual basis (ω^j).

If the basis is changed from (e_i) to $(e_{i'})$ by means of the non-singular linear transformation

$$e_{i'} = \sum_{i=1}^{m} p^i_{i'} e_i ,$$

then

$$\omega^{i'} = \sum_{i=1}^{m} p^{i'}{}_i \omega^i \text{ where } (p^i_{i'}) \text{ and } (p^{i'}_i) \text{ are}$$

inverse matrices. Then

$$T^{i_1' \ldots i_r'}_{j_1' \ldots j_s'} = T(\omega^{i_1'}, \omega^{i_2'} \ldots, \omega^{i_r'}, e_{j_1'}, e_{j_2'}, \ldots e_{j_s'})$$

$$= p^{i_1'}_{i_1} p^{i_2'}_{i_2} \ldots p^{i_r'}_{i_r} p^{j_1}_{j_1'} p^{j_2}_{j_2'} \ldots p^{j_s}_{j_s'} T^{i_1 i_2 \ldots i_r}_{j_1 j_2 \ldots j_s} \tag{1.1}$$

Here we use the Einstein convention of summing over repeated suffixes, to avoid numerous Σ signs. This gives the classical law of transformation of components of a tensor T of type (r, s) when the basis is changed.

Let S be a tensor of type (p, q) and T a tensor of type (r, s). Then $S \otimes T$ can be defined as the tensor whose components relative to a basis (e_i) are

$$S^{i_1 \ldots i_p}_{j_1 \ldots j_q} T^{m_1 \ldots m_r}_{n_1 \ldots n_s} .$$

We leave it as an exercise to give an invariant definition of $S \otimes T$.

1.4.1 Contraction of tensors of type (r, s) when $r \geqslant 1$, $s \geqslant 1$

This is an algebraic operation for defining new tensors from old ones. Suppose we are given T with components $T^{i_1 \ldots i_r}_{j_1 \ldots j_s}$. Then by equating one upper suffix with one lower suffix and summing according to the Einstein convention, we get a new tensor of type $(r-1, s-1)$. Obviously there are many possible ways of contracting to form a new tensor if $r > 1$ and $s > 1$.

For example, a tensor of type $(r-1, s-1)$ is determined by $T^{i_1 i_2 i_3 \ldots i_r}_{j_1 i_1 j_3 \ldots j_s}$ obtained by equating j_2 with i_1 and summing. Clearly the tensor product of r tensors of type $(1, 0)$ and s tensors $(0, 1)$ is a tensor of type (r, s). In general it is impossible to write a given tensor of type (r, s) as a product of tensors of types $(1, 0)$ and $(0, 1)$, but it is always possible to express it as a finite sum of such tensors. This follows because a basis of tensors of type (r, s) is given by

$$e_{i_1}\otimes e_{i_2}\otimes\ldots\otimes e_{i_r}\otimes\omega^{j_1}\otimes\omega^{j_2}\otimes\ldots\otimes\omega^{j_s}.$$

Consider now any decomposable tensor of type (r, s) which we may write as

$$X_1\otimes X_2\otimes\ldots\otimes X_r\otimes\eta^1\otimes\eta^2\otimes\ldots\otimes\eta^s,$$

where $X_i\in\mathscr{D}^1(M)$ and $\eta^j\in\mathscr{D}_1(M)$.

The operation of contracting i_1 with j_2 is now seen to be the evaluation of η^2 acting on X_1 to give the C^∞-function $\eta^2(X_1)$. The contracted tensor is now

$$\eta^2(X_1)X_2\otimes\ldots\otimes X_r\otimes\eta^1\otimes\eta^3\otimes\ldots\otimes\eta^s.$$

Since any tensor is expressible as a linear combination of decomposable tensors, this process of contraction can be extended to an arbitrary tensor in a coordinate free way. However, in practice, contraction is applied to components of tensors, and that is why we have defined contraction in terms of components.

1.5 SPECIAL TENSORS

A special tensor of type (1, 1) is the so-called Kronecker delta, with components referred to some basis of V given by δ^i_j where $\delta^i_j=1$ if $i=j$, $\delta^i_j=0$ otherwise. Since

$$\delta^{i'}_{j'}=p^{i'}_i p^j_{j'}\delta^i_j=p^{i'}_j p^j_{j'}\ ,$$

it follows that our definition is valid relative to any basis of V.

There is a natural isomorphism between tensors of type (1, 1) and linear transformations of V. Under this isomorphism the Kronecker delta corresponds to the identity transformation. For, if $\{e_i\}$, $i=1, 2\ldots, n$, is a basis for V, then a tensor of type (1, 1) is specified by its components relative to $\{e_i\}$, say (t^i_j). This matrix determines uniquely a linear transformation of V. Conversely, a linear transformation of V determines a matrix (t^i_j) relative to a basis, and this in turn determines a tensor of type (1, 1). Similarly a real number can be regarded as a tensor of type (0, 0).

A tensor T of type (r, s) is said to be symmetric with respect to the places k and l if it has the same value when the kth and lth arguments are interchanged, it being understood that the arguments are of the same type, that is, both are ω's or both are e's. In a similar way we may consider skew-symmetric tensors. In particular we shall be concerned with tensors of type $(0, s)$ which are skew-symmetric with respect to all places—these are called **alternating tensors**.

Clearly the space of alternating tensors of type $(0, s)$ form a vector subspace of dimension $\binom{m}{s}$, where $s \leqslant m$. In particular when $s=m$ we note that the subspace is 1-dimensional, and that the dimension is 0 when $s>m$. The set of all alternating tensors can be made into an algebra, called the **Grassmann algebra** of V. The product operation (denoted by $\wedge$) is defined as follows:

Denote by $\mathscr{A}_r$ the space of alternating tensors of type $(0, r)$. Let $f, g \in \mathscr{A}_0$; $\theta \in \mathscr{A}_r$, $\omega \in \mathscr{A}_s$; $X_i \in V$. Then

$$f \wedge g = fg,$$

$$(f \wedge \theta)(X_1, \ldots, X_r) = f\theta(X_1, \ldots, X_r),$$

$$(\omega \wedge g)(X_1, \ldots, X_s) = g\omega(X_1, \ldots, X_s),$$

$$(\theta \wedge \omega)(X_1, \ldots, X_{r+s}) = \frac{1}{(r+s)!} \sum_{\sigma \in S_{r+s}} \epsilon(\sigma)\theta(X_{\sigma(1)}, \ldots, X_{\sigma(r)}).$$

$$\omega(X_{\sigma(r+1)}, \ldots, X_{\sigma(r+s)}),$$

where S_{r+s} is the symmetric group of permutations on $\{1, 2, \ldots, r+s\}$ and $\epsilon(\sigma)$ is the sign of the permutation σ. For example if $\theta \in \mathscr{A}_2$, $\omega \in \mathscr{A}_1$, then

$$(\theta \wedge \omega)(X_1, X_2, X_3) = \frac{1}{3!}\begin{cases} \theta(X_1, X_2)\omega(X_3) - \theta(X_2, X_1)\omega(X_3) \\ +\theta(X_2, X_3)\omega(X_1) - \theta(X_3, X_2)\omega(X_1) \\ +\theta(X_3, X_1)\omega(X_2) - \theta(X_1, X_3)\omega(X_2) \end{cases}$$

$$= \frac{1}{3}\{\theta(X_1, X_2)\omega(X_3) + \theta(X_2, X_3)\omega(X_1) + \theta(X_3, X_1)\omega(X_2)\}.$$

Evidently $\theta \wedge \omega$ is the alternating part of the tensor product $\theta \otimes \omega$. From the definition of $\wedge$, we have

$$\theta \wedge \omega = (-1)^{rs}\omega \wedge \theta.$$

Let $\theta^1, \theta^2, \ldots, \theta^l$ be l elements of $\mathscr{A}_1$, that is, elements of the dual space V^*, and let

$$\omega^i = \sum_{j=1}^{l} a_{ij}\theta^j,\ 1 \leqslant i, j \leqslant l,\ (a_{ij} \in \mathbb{R}).$$

Then it follows that

$$\omega^1 \wedge \omega^2 \wedge \ldots \wedge \omega^l = \det(a_{ij})\theta^1 \wedge \theta^2 \wedge \ldots \wedge \theta^l.$$

Indeed, Grassmann invented his algebra primarily to deal with determinants and the theory of linear dependence in vector spaces.

Another example of a special tensor arises when the vector space V is given an inner product structure $\langle\,,\,\rangle$. This is a bilinear mapping $V \times V \rightarrow \mathbb{R}$ which satisfies

(i) $<v, w> = <w, v>$,
(ii) $<v, w> = 0$ for all $v \in V \Rightarrow w = 0$.

If we write $<e_i, e_j> = g_{ij}$, then clearly the inner product $<, >$ determines a symmetric tensor of order (0, 2). Requirement (ii) implies that $\det(g_{ij}) \neq 0$.

1.6 FIBRE BUNDLES

The theory of fibre bundles and its applications has played a very significant role in the development of mathematics since 1940. Although it has now developed into a branch of mathematics worthy of study in its own right, it originated in differential geometry in the work of Whitney and Stiefel. Moreover, the important paper by Weyl (1939) already contained the essential ideas of characteristic classes. The first textbook on the subject was written by Steenrod (1951). A more up to date treatment is given in the book by Husemoller (1966).

We shall be primarily interested in the tangent bundle and the frame bundle of differentiable manifolds, and in the tangent sphere bundle of Riemannian manifolds. More generally, the tensors of type (r, s) over a differentiable manifold form a fibre bundle. We shall consider the special case of tensors of type (0, 2), primarily to illustrate the essential properties of a fibre bundle. The reader will have little difficulty in extending the ideas to apply to other fibre bundles.

Let $\tilde{V}$ denote a real m-dimensional vector space, equipped with a fixed basis (e_i), $i = 1, 2, \ldots, m$. This determines a dual basis (ω^i) in the dual space of $\tilde{V}$, and a uniquely determined basis $(\omega^i \otimes \omega^j)$ in the space $\tilde{V}_2^0$ of tensors of type (0, 2).

Let M be a differentiable manifold and let U be a coordinate neighbourhood equipped with coordinates x_i, $i = 1, 2, \ldots, m$. Then at a point $p \in U$ there is a coordinate basis $(\partial/\partial x_i)_p$ for the tangent space at p, and a corresponding coordinate basis $dx_i \otimes dx_j$ for the space $T_2^0(p)$ of tensors of type (0, 2) at p. The fibre bundle which we shall construct will have as its bundle space all the possible tensors of type (0, 2) at all points $p \in M$. We first restrict p to U and consider the mapping

$$\phi_U \colon U \times \tilde{V}_2^0 \to B_U$$

where $B_U = \bigcup_{p \in U} T_2^0(p)$, specified by requiring that $\phi_U(p, y)$, for $y \in \tilde{V}_2^0$, belongs to $T_2^0(p)$ and has the same components relative to the basis $(dx_i \otimes dx_j)_p$ as y has relative to the basis $(\omega^i \otimes \omega^j)$ of $\tilde{V}_2^0$.

It is convenient to introduce another mapping for fixed p

$$\Phi_{U,p} \colon \tilde{V}_2^0 \to T_2^0(p)$$

given by

$$\phi_{U,p}(y)=\phi_U(p, y).$$

Now let V be another coordinate neighbourhood such that $U \cap V \neq \varnothing$ and let $\phi_V, \phi_{V,p}$ be maps associated with V. We denote by $g_{UV}(p)$ the map defined for $p \in U \cap V$ by $g_{UV}(p)$: $\tilde{V}_2^0 \to \tilde{V}_2^0$,

$$g_{UV}(p)=\phi_{U,p}^{-1} \circ \phi_{V,p} \; .$$

The inverse map $\phi_{U,p}^{-1}$ exists because ϕ_U is clearly injective. We now obtain an explicit formula for $g_{UV}(p)$,

Let $y \in \tilde{V}_2^0$ have components y_{ij} so that

$$y=\sum_{i,j} y_{ij}\omega^i \otimes \omega^j \; .$$

Write

$$y'=g_{UV}(p)y$$

with components y_{ij}' so that

$$y'=\sum_{i,j} y_{ij}'\omega^i \otimes \omega^j \; .$$

Then we must have

$$\phi_U(p, y')=\phi_V(p, y) \; ,$$

$$\sum_{i,j} y_{ij}' du^i \otimes du^j=\sum_{i,j} y_{ij} dv^i \otimes dv^j \; .$$

Since

$$dv^i=\sum_j \left(\frac{\partial v^i}{\partial u^j}\right)_{U(p)} du^j$$

this gives

$$y_{ij}'=\sum_{r,s} \left(\frac{\partial v^r}{\partial u^i}\right)_{U(p)} \left(\frac{dv^s}{\partial u^j}\right)_{U(p)} y_{rs} \; .$$

Thus $g_{UV}(p)$ is a linear automorphism of $\tilde{V}_2^0$. We give to $\tilde{V}_2^0$ the topology and the C^∞-differentiability of the euclidean space of the components of its elements. Denote by $GL(\tilde{V}_2^0)$ the general linear group in $\tilde{V}_2^0$. Then $g_{UV}(p)$ defines a mapping g_{UV}: $U \cap V \to GL(\tilde{V}_2^0)$ which is C^∞ in $U \cap V$.

Now take a covering of M by coordinate neighbourhoods $U, V, W, \ldots$, and denote the corresponding maps by $\phi_U, \phi_V, \phi_W, \ldots$. Since the maps $g_{UV}, g_{VW}, \ldots$ are C^∞ (and *a fortiori* continuous) it follows from Steenrod

(1951), page 14, that the set of all tensors of type (0, 2) form a fibre bundle with ϕ_U as coordinate functions. The bundle space B is given the compact open topology, that is, ϕ_U maps open sets of $U \cap \tilde{V}_2^0$ into open sets of the bundle space B.

Denote by $\pi: B \rightarrow M$, the projection map

$$\pi(T_2^0(p)) = p.$$

A map $s: M \rightarrow B$ such that $\pi \circ s =$ identity is called a cross-section. Thus a cross-section of the bundle B is a tensor field of type (0, 2) over M. The question of the existence of a cross-section with special properties is very important in global differential geometry. For example, the existence of a nowhere zero vector field imposes severe topological restrictions on M. However, a standard theorem in fibre bundle theory asserts that if the type fibre is homeomorphic to a euclidean cell, then the existence of a cross-section of the bundle is guaranteed. If instead of considering the tensors of type (0, 2), we had considered the special case of positive definite symmetric tensors of type (0, 2) we would still have obtained a fibre bundle whose type fibre [a linear subspace of $\tilde{V}_2^0$] is homeomorphic to a cell. This implies a fundamental theorem—namely, the existence of a Riemannian metric on a differentiable manifold M.

Incidentally the maps g_{UV} corresponding to a covering of M generate a group which may be a proper subgroup of the general linear group. A natural problem is to choose a covering such that the corresponding group generated by g_{UV} is as small as possible. We return to this later.

1.6.1 The frame bundle

A **frame** consists of a point $p \in M$ and m linearly independent tangent vectors at p. The set of all frames forms a principal bundle B called the frame bundle—the dimension of the bundle space being $m + m^2$. Let U be a coordinate neighbourhood of M, with local coordinates (u^i). Then there corresponds a local coordinate system (u^i, X_i^k) for B determined by the m vectors of the frame which are given by

$$X_i = \sum_k X_i^k \frac{\partial}{\partial u^k},$$

Again, the group generated by g_{UV} is the full linear group $GL(m)$ or perhaps a proper subgroup,

If we have a Riemannian metric we can consider a sub-bundle consisting of orthonormal frames—that is, each of the m vectors has unit length and the vectors are mutually orthogonal. The corresponding group generated by g_{UV} in this case is the orthogonal group $O(m)$ or a proper subgroup. We could define the Riemannian manifold M as **orientable** if this group is reducible to the special orthogonal group $SO(m)$,

Similarly we can consider the bundle of unit tangent vectors, known as the **unit sphere bundle**; again the corresponding group is $O(m)$.

1.7 EXTERIOR DIFFERENTIAL FORMS

An exterior differential form of degree r on M is a law which associates with each point $p \in M$ an element of $\mathscr{A}_r$ of the Grassmann algebra of M_p. As in the case of vector fields we want to restrict the differential forms to be C^∞. This is automatically achieved if we define the space of exterior differential r-forms $\mathscr{A}_r(M)$ to be the set of alternate $C^\infty(M)$-multilinear mappings of $\mathscr{D}^1 \times \mathscr{D}^1 \times \ldots \times \mathscr{D}^1$ (r-times) into $C^\infty(M)$. All the algebraic properties of the Grassmann algebra now pass over to exterior forms. In particular, equations satisfied by the alternating tensors in section 1.5 are still valid for exterior differential forms, but here we take $f, g \in C^\infty(M)$, $\theta \in \mathscr{A}_r(M)$, $\omega \in \mathscr{A}_s(M)$, $X_i \in \mathscr{D}^1(M)$. We denote by $\mathscr{A}(M) = \sum_{s=0}^{\infty} \mathscr{A}_s(M)$.

We may associate with the Grassmann algebra of exterior forms an operation d called **exterior differentiation** which maps $\mathscr{A}_s$ into $\mathscr{A}_{s+1}$. The importance of this operation is that it is intrinsic to the structure, that is, no other extra structure is necessary for its definition other than the differentiable structure on M. We shall see later that this contrasts strongly with what happens if we consider tensor fields in general.

At first sight it appears that the following definition of d is unnecessarily complicated. The reason for this is that although it is easy to define a suitable operator for each coordinate neighbourhood of M, we want to use our operator over the whole of M. Our treatment follows closely that of Helgason:

Theorem 3. *There exists a unique $\mathbb{R}$-linear mapping d; $\mathscr{A}(M) \to \mathscr{A}(M)$ such that:*

(*i*) $d\mathscr{A}_s(M) \subset \mathscr{A}_{s+1}(M)$ *for each* $s \geqslant 0$.
(*ii*) *If* $f \in C^\infty(M)$, *then* df *is the* 1-*form given by* $df(X) = Xf$, $X \in \mathscr{D}^1(M)$.
(*iii*) $d \circ d = 0$,
(*iv*) $d(\omega_1 \wedge \omega_2) = d\omega_1 \wedge \omega_2 + (-1)^r \omega_1 \wedge d\omega_2$ if $\omega_1 \in \mathscr{A}_r(M)$, $\omega_2 \in \mathscr{A}(M)$.

The factor $(-1)^r$ may seem strange at first sight, but we shall soon see that it is appropriate.

We have to prove both uniqueness and existence. We assume first that such an operation exists and prove that it is necessarily unique. On the way we obtain a formula for d, and by using this formula as the definition of d, we prove existence.

Let $p \in M$ and let U be a coordinate neighbourhood of p with coordinates $\{x_1, \ldots, x_m\}$. Let V be an open subset of U such that $\bar{V}$ is compact and $p \in V$, $\bar{V} \subset U$. From (ii) we see by taking $f = x_i$, that the forms dx_i $(1 \leqslant i \leqslant m)$ on U satisfy $dx_i\left(\frac{\partial}{\partial x^j}\right) = \delta_{ij}$ on U. This justifies our notation in section 1.32. Moreover, it follows that

$$dx_{i_1} \wedge dx_{i_2} \wedge \ldots \wedge dx_{i_r}$$

is a basis for the r-forms over U. Let $\theta \in \mathscr{A}(M)$, and denote by θ_U the restriction of θ to U. Then we can write

$$\theta_U = \Sigma f_{i_1 i_2 \ldots i_r} dx_{i_1} \wedge dx_{i_2} \wedge \ldots \wedge dx_{i_r}, \quad f_{i_1 i_2 \ldots i_r} \in C^\infty(U),$$

and we call this the **expression of** θ_U **on** U. To establish consistency, we next prove that for any θ,

$$d(\theta_V) = (d\theta)_V .$$

We now make use of the extension Theorem 1. It follows that there exist functions $\psi_{i_1 \ldots i_r} \in C^\infty(M)$, $\phi_i \in C^\infty(M)$ $(1 \leqslant i \leqslant m)$ such that

$$\psi_{i_1 \ldots i_r} = f_{i_1 \ldots i_r}, \ \phi_1 = x_1, \ \phi_2 = x_2, \ldots, \phi_m = x_m \text{ on } V.$$

Write $\quad \omega = \psi_{i_1 \ldots i_r} \, d\phi_{i_1} \wedge \, d\phi_{i_2} \wedge \ldots \wedge d\phi_{i_r}$.

We want to prove that $(d\theta)_V = (d\omega)_V$. Well, clearly $\theta_V = \omega_V$. Moreover, from (iv) we have

$$d(f(\theta - \omega)) = df \wedge (\theta - \omega) + f d(\theta - \omega)$$

for all $f \in C^\infty(M)$.

We may choose for f a C^∞-function which is identically 0 outside V, and is identically 1 on an open subset of V. From this we get $(d\theta)_V = (d\omega)_V$ as required.

Since

$$d\omega = \Sigma d\psi_{i_1 \ldots i_r} \wedge d\phi_{i_1} \wedge \ldots \wedge d\phi_{i_r}$$

(from (iii) and (iv), Theorem 3) and since

$d(f_V) = (df)_V$ for each $f \in C^\infty(M)$, we conclude that

$$(d\omega)_V = \Sigma \, df_{i_1 \ldots i_r} \wedge dx_{i_1} \wedge dx_{i_2} \wedge \ldots \wedge dx_{i_r} .$$

Thus we have proved the relation

$$(d\theta)_V = d(\theta_V) = \Sigma df_{i_1 \ldots i_r} \wedge dx_{i_1} \wedge dx_{i_2} \wedge \ldots \wedge dx_{i_r}, \tag{1.2}$$

which clearly establishes the uniqueness on V.

On the manifold M itself we have the formula

$$(p+1) d\omega(X_1, \ldots, X_{p+1}) = \sum_{i=1}^{p+1} (-1)^{i+1} X_i(\omega(X_1, \ldots, \hat{X}_i, \ldots, X_{p+1}) + \sum_{i<j} (-1)^{i+j} \omega([X_i, X_j], X_1, \ldots, \hat{X}_i, \ldots, \hat{X}_j, \ldots, X_{p+1}), \tag{1.3}$$

for $\omega \in \mathscr{A}_p(M)$ $(p \geqslant 1)$, $X_i \in \mathscr{D}^1(M)$.

To prove the above formula, we merely have to check that it holds in a coordinate neighbourhood of each point—in fact, the difficulty of this analysis is in establishing this very fact. The actual verification is a straightforward consequence of (1.2).

Now to prove the *existence* of d, we may take (1.3) together with property (ii) as the *definition* of d. It is not difficult to prove that properties (i), (iii) and (iv) are satisfied, and so the main theorem is proved.

The reader may be puzzled why our definition of d is more complicated than the definition given in several other books. The reason is that if you are working in a single coordinate neighbourhood, then the consistency question does not arise. The reader is also warned that we use the term $(p+1)$ in (1.3) whereas some authors omit this. For example, when $p=1$, our formula gives

$$2d\omega(X_1, X_2) = X_1(\omega(X_2)) - X_2(\omega(X_1)) - \omega([X_1, X_2]),$$

whereas the corresponding formula from O'Neill (1966), p. 158, omits the factor 2 on the left-hand side. We do, however, agree with O'Neill's statement on p. 155—"It is difficult to exaggerate the importance of the exterior derivative".

1.8 TENSOR FIELDS

A tensor field T of type (r, s) over M is a law which associates with each point $p \in M$ a tensor of type (r, s) with respect to the m-dimensional tangent space at p.

T is of class C^∞ at p if the components of T relative to some coordinate basis in a neighbourhood containing p are C^∞-functions. T is of class C^∞ over M if T is of class C^∞ at p for every $p \in M$.

We denote by $\mathscr{D}^r_s(M)$ the space of C^∞ tensor fields over M of type (r, s). This is consistent with our notation of $\mathscr{D}^1(M)$ for the space of vector fields (tensor fields of type (1,0)) and of $\mathscr{D}_1(M)$ for the space of 1-forms over M.

Clearly an alternative definition of a tensor field of type (r, s) is a multilinear mapping of $(\mathscr{D}_1(M) \times \mathscr{D}_1(M) \times \ldots \times \mathscr{D}_1(M)) \times (\mathscr{D}^1(M) \times \ldots \times \mathscr{D}^1(M))$ into $C^\infty(M)$, where the first factor contains r terms and the second s terms. Moreover, multilinearity is with respect to the ring of C^∞-functions. In fact, the set of all tensor fields forms a module over the ring $C^\infty(M)$.

Suppose we have a tensor field defined over some coordinate neighbourhood U with coordinates x_i, and some intersecting coordinate neighbourhood U' with coordinates $x_{i'}$. Then the components of T with respect to the coordinate bases $\frac{\partial}{\partial x_i}, \frac{\partial}{\partial x_{i'}}$ are related by (1.1) of section 1.4 where $p^i_{i'} = \left(\frac{\partial x_i}{\partial x_{i'}}\right)$, $p^{i'}_i = \left(\frac{\partial x_{i'}}{\partial x_i}\right)$. In fact, this is the way in which many older books on differential geometry defined a tensor, that is, by specifying the "law of transformation" of components associated with a change of coordinates.

We have already seen that the module of exterior differential forms possesses an intrinsically determined operation of exterior differentiation. However, the module of arbitrary tensor fields does not possess such an operator. If we wish to perform some analogous form of differentiation on tensor fields, it is necessary to introduce a new geometrical object called an affine connexion. We do this in the next section.

1.9 AFFINE CONNEXION

We introduce a differential operator which acts on vector fields and functions, and finally we extend its domain of operation to arbitrary tensor fields.

An **affine connexion** on a manifold M is a rule ∇ which associates to each $X \in \mathscr{D}^1(M)$ a linear mapping ∇_X of the space $\mathscr{D}^1(M)$ into itself such that

$$(\nabla_1) \qquad \nabla_{fX+gY} = f\nabla_X + g\nabla_Y \ ;$$

$$(\nabla_2) \qquad \nabla_X(fY) = f\nabla_X Y + (Xf)Y$$

for $f, g \in C^\infty(M)$; $X, Y \in \mathscr{D}^1(M)$.

The operator ∇_X is called **covariant differentiation** with respect to X.

An affine connexion ∇ on M induces an affine connexion ∇_U on a coordinate neighbourhood U. Suppose the coordinate system of U is given by ϕ: $q \to (x_1(q), \ldots, x_m(q))$. Write ∇_i instead of $(\nabla_U)_{\frac{\partial}{\partial x_i}}$.

Then the equation

$$\nabla_i\left(\frac{\partial}{\partial x_j}\right) = \sum_k \Gamma^k_{ji} \frac{\partial}{\partial x^k}$$

defines n^3 C^∞-functions Γ^k_{ji} on U. These are called **connexion coefficients**. Suppose V is an overlapping coordinate neighbourhood with corresponding coordinates $x_{i'}$, and write

$$\nabla_{i'}\left(\frac{\partial}{\partial x_{j'}}\right) = \sum_{k'} \Gamma^{k'}_{j'i'} \frac{\partial}{\partial x_{k'}} \ .$$

Then axioms (∇_1) and (∇_2) imply

$$\Gamma^{k'}_{i'j'} = \sum_{i,j,k} \frac{\partial x_i}{\partial x_{i'}} \frac{\partial x_j}{\partial x_{j'}} \frac{\partial x_{k'}}{\partial x_k} \Gamma^k_{ij} + \sum_j \frac{\partial^2 x_j}{\partial x_{i'} \partial x_{j'}} \frac{\partial x_{k'}}{\partial x_j} \ .$$

Again, some older books define a connexion as a set of n^3 real-valued functions relative to some coordinate system (x_i). When the coordinates are changed from x_i to $x_{i'}$, the corresponding connexion coefficients are given by the above equation. Of course, one then has to verify that a further transformation of coordinates $x_{i'} \to x_{i''}$ still gives consistent results—this is unnecessary with our definition.

It is convenient to extend the action of ∇_X to C^∞-functions by writing $\nabla_X f = Xf$. Then (∇_2) may be written

$$(\nabla_2)' \qquad \nabla_X(fY) = f\nabla_X Y + \nabla_X f . Y \ ,$$

which we may regard as analogous to the Leibnitz rule for differentiating a product.

We may extend the definition of ∇_X to arbitrary tensor fields. For example, let us consider how to extend ∇_X to a tensor field g of type (0,2). It is reasonable to write

$$\nabla_X\{g(Y, Z)\} = (\nabla_X g)(Y, Z) + g(\nabla_X Y, Z) + g(Y, \nabla_X Z) \ .$$

This gives the required extension, namely,

$$(\nabla_X g)(Y, Z) = X\{g(Y, Z)\} - g(\nabla_X Y, Z) - g(Y, \nabla_X Z) \ .$$

If we write $X = \dfrac{\partial}{\partial x_i}$, $Y = \dfrac{\partial}{\partial x_j}$, $Z = \dfrac{\partial}{\partial x_k}$, this gives

$$(\nabla_i g)_{jk} = \frac{\partial}{\partial x_i} g_{jk} - g_{pk}\, \Gamma^p_{ji} - g_{jp}\, \Gamma^p_{ki}$$

where we have used the summation convention to simplify the notation.

For a tensor field ω, of type (0, 1), we write

$$\nabla_X\{\omega(Y)\} = (\nabla_X \omega) Y + \omega\{\nabla_X Y\}, \text{ giving}$$

$$(\nabla_X \omega) Y = X\{\omega(Y)\} - \omega(\nabla_X Y) \ .$$

In terms of suffixes this is equivalent to

$$(\nabla_i \omega)_j = \frac{\partial}{\partial x_i} \omega_j - \omega_p\, \Gamma^p_{ji} \ .$$

In a similar manner we derive the relation for a tensor T of type (r,s)

$$\nabla_i\, T^{j_1 \dots j_r}_{k_1 \dots k_s} = \frac{\partial}{\partial x_i}\, T^{j_1 \dots j_r}_{k_1 \dots k_s} + T^{p j_2 \dots j_r}_{k_1 \dots k_s} \Gamma^{j_1}_{pi} + T^{j_1 p j_3 \dots j_r}_{k_1 \dots k_s}\, \Gamma^{j_2}_{pi} + \dots$$

$$- T^{j_1 \dots j_r}_{p k_2 \dots k_s}\, \Gamma^p_{k_1 i} - T^{j_1 \dots j_r}_{k_1 p k_3 \dots k_s}\, \Gamma^p_{k_2 i} \ \dots,$$

where on the right there is a + term for each upper suffix and a − term for each lower suffix.

1.10 THE STRUCTURAL EQUATIONS

1.10.1 The torsion and curvature tensors

Let M be a manifold with a connexion ∇. We define as follows two tensors associated with ∇, namely the **torsion** tensor T and the **curvature** tensor R.

$$T(X, Y) = \nabla_X(Y) - \nabla_Y(X) - [X, Y] ,$$
$$R(X, Y) = \nabla_{[X, Y]} - \nabla_X \nabla_Y + \nabla_Y \nabla_X$$

for all $X, Y \in \mathscr{D}^1(M)$.

Note that our definition of $R(X, Y)$ has the opposite sign from that in Helgason's book. Clearly

$$\begin{aligned} T(fX, gY) &= \nabla_{fX}(gY) - \nabla_{gY}(fX) - [fX, gY] \\ &= f(Xg)Y + fg\nabla_X Y - g(Yf)X - fg\nabla_Y X - f(Xg)Y + g(Yf)X - \\ &\qquad fg\,[X, Y] \\ &= fg\,[\nabla_X(Y) - \nabla_Y(X) - [X, Y]] \\ &= fgT(X, Y), \text{ where } X, Y \in \mathscr{D}^1(M); f, g \in C^\infty(M) . \end{aligned}$$

1.10.2 Exercises

(1) Prove that
$R(fX, gY).hZ = fghR(X, Y)Z$ for all $X, Y, Z \in \mathscr{D}^1(M); f, g, h \in C^\infty(M)$.

(2) Prove that T is a bilinear mapping of $\mathscr{D}^1(M) \times \mathscr{D}^1(M)$ into $\mathscr{D}^1(M)$ with respect to $C^\infty(M)$. Show that such mappings are isomorphic to the space of tensors $\mathscr{D}^1_2$. Hence show that the torsion is a tensor field of type (1, 2).

(3) Prove that R is a bilinear mapping of $\mathscr{D}^1(M) \times \mathscr{D}^1(M) \times \mathscr{D}^1(M)$ into $\mathscr{D}^1(M)$ with respect to $C^\infty(M)$. Hence show that the curvature is a tensor field of type (1, 3).

1.10.3 Cartan's equations

Let $X_1, X_2, \ldots, X_m$ be a basis of vector fields over some open set U of M, and let $\omega^1, \omega^2, \ldots, \omega^m$ be the dual basis. We define functions Γ^k_{ji}, T^k_{ij}, R^k_{lij} over U by the formulas

$$\nabla_{X_i}(X_j) = \Gamma^k_{ji} X_k ,$$
$$T(X_i, X_j) = T^k_{ij} X_k ,$$
$$R(X_i, X_j).X_l = R^k_{lij} X_k$$

where we have used the summation convention.

Let ω^i_j $(1 \leqslant i, j \leqslant m)$ be a matrix of 1-forms on U given by

$$\omega^i_j = \Gamma^i_{jk} \omega^k .$$

Clearly the matrix (ω^i_j) determines the functions Γ^i_{jk} and hence the connexion. We now show that this matrix determines the torsion and curvature fields. More precisely, we establish the following structural equations of É. Cartan:

$$d\omega^i + \omega^i_p \wedge \omega^p = \tfrac{1}{2}T^i_{jk}\omega^j \wedge \omega^k \ , \tag{1.4}$$

$$d\omega^i_l + \omega^i_p \wedge \omega^p_l = -\tfrac{1}{2}R^i_{ljk}\omega^j \wedge \omega^k \tag{1.5}$$

To verify these equations, it is sufficient to prove that the 2-forms on each side give the same value to an arbitrary pair of basis vector fields X_j, X_k.

The left hand side of (1.4) gives

$$\begin{aligned} d\omega^i(X_j, X_k) + \omega^i_p \wedge \omega^p(X_j, X_k) = {} & \tfrac{1}{2}\{X_j\omega^i(X_k) - X_k\omega^i(X_j) - \omega^i[X_j, X_k]\} \\ & + \tfrac{1}{2}\{\omega^i_p(X_j)\omega^p(X_k) - \omega^i_p(X_k)\omega^p(X_j)\} \ . \end{aligned}$$

If we define functions c^i_{jk} by $[X_j, X_k] = c^i_{jk}X_i$, this becomes

$$-\tfrac{1}{2}c^i_{jk} + \tfrac{1}{2}(\Gamma^i_{kj} - \Gamma^i_{jk}) \ .$$

The right hand side of (1.4) gives

$$\begin{aligned} \tfrac{1}{2}T^i_{pq}\omega^p \wedge \omega^q(X_j, X_k) &= \tfrac{1}{4}T^i_{pq}\{\delta^p_j\delta^q_k - \delta^p_k\delta^p_j\} \\ &= \tfrac{1}{2}T^i_{jk} \ . \end{aligned}$$

But

$$\begin{aligned} T(X_j, X_k) &= \nabla_{X_j}(X_k) - \nabla_{X_k}(X_j) - [X_j, X_k] \\ &= (\Gamma^i_{kj} - \Gamma^i_{jk} - c^i_{jk})X_i \\ &= T^i_{jk}X_i \ , \end{aligned}$$

so we have shown that the left-hand side of (1.4) is equal to the right-hand side.

Exercise

Using the above method establish the validity of (1.5)

Solution

The left-hand side of (1.5) gives

$$\begin{aligned} & (d\omega^i_l + \omega^i_p \wedge \omega^p_l)(X_j, X_k) \\ & \quad = \tfrac{1}{2}\{X_j(\omega^i_l(X_k)) - X_k(\omega^i_l(X_j)) - \omega^i_l([X_j, X_k])\} \\ & \qquad + \tfrac{1}{2}\{\omega^i_p(X_j)\omega^p_l(X_k) - \omega^i_p(X_k)\omega^p_l(X_j)\} \\ & \quad = \tfrac{1}{2}\{X_j(\Gamma^i_{lk}) - X_k(\Gamma^i_{lj}) - \Gamma^i_{lp}c^p_{jk} + \Gamma^i_{pj}\Gamma^p_{lk} - \Gamma^i_{pk}\Gamma^p_{lj}\} \ . \end{aligned}$$

The right-hand side of (1.5) gives

$$
\begin{aligned}
-\tfrac{1}{2}R^i_{lpq}\omega^p\wedge\omega^q(X_j, X_k) \\
&= -\tfrac{1}{4}R^i_{lpq}(\delta^p_j\delta^q_k - \delta^p_k\delta^q_j) \\
&= -\tfrac{1}{2}R^i_{ljk}\ .
\end{aligned}
$$

But

$$
\begin{aligned}
R(X_j, X_k).X_l = R^i_{ljk}X_i \\
&= \nabla_{[X_j,X_k]}X_l - \nabla_{X_j}(\nabla_{X_k}(X_l)) + \nabla_{X_k}(\nabla_{X_j}(X_l)) \\
&= c^p_{jk}\Gamma^i_{lp}X_i - \nabla_{X_j}(\Gamma^p_{lk}X_p) + \nabla_{X_k}(\Gamma^p_{lj}X_p) \\
&= (c^p_{jk}\Gamma^i_{lp} - X_j(\Gamma^i_{lk}) - \Gamma^p_{lk}\Gamma^i_{pj} + X_k(\Gamma^i_{lj}) + \Gamma^p_{lj}\Gamma^i_{pk})X_i.
\end{aligned}
$$

It follows that (1.5) is established.

We shall make great use of the structural equations later in the book.

1.11 MAPPING OF MANIFOLDS

Let M and N be C^∞-manifolds of dimension m and n respectively. Let Φ: $M\to N$ be a mapping of M into N. The mapping is said to be differentiable at $p\in M$ if $g\circ\Phi\in C^\infty(p)$ for every $g\in C^\infty(\Phi(p))$. The mapping Φ is called **differentiable** if it is differentiable at p for all $p\in M$.

Let $(x_1, x_2, \ldots, x_m)$ be coordinates over a neighbourhood U of p and let $(y_1, y_2, \ldots, y_n)$ be coordinates over a neighbourhood V of $\Phi(p)$ such that $\Phi(U)\subset V$. If ψ, ψ' are the chart maps associated with U and V, then the mapping $\psi'\circ\Phi\circ\psi^{-1}$ of $\psi(U)$ into $\psi'(V)$ is given by a system of n functions

$$y_j = \phi_j(x_1, x_2, \ldots, x_m) \quad (1\leqslant j\leqslant n)\ . \tag{1.6}$$

Clearly Φ is differentiable at p if and only if each of the real-valued functions ϕ_j have partial derivatives of all orders in some neighbourhood of $\psi(p)$. In an analogous way we may define an analytic mapping of analytic manifolds.

The mapping Φ is called a diffeomorphism of M onto N if Φ is a bijective differentiable map of M onto N and Φ^{-1} is differentiable. In particular we may consider the set of diffeomorphisms of M onto itself.

If Φ: $M\to N$ is differentiable. then at each point $p\in M$ there is induced a very important linear map called the **differential** of Φ, denoted by Φ_*, which maps tangent vectors at p to tangent vectors at $\Phi(p)$. Let $X\in M_p$, the tangent space to M at p. Consider the mapping $C^\infty(\Phi(p))\to\mathbb{R}$ given by

$$\Phi_*(X)(g) = X(g\circ\Phi)\ .$$

Clearly $\Phi_*(X)$ has the required properties of a tangent vector at $\Phi(p)$.

We now show that relative to coordinate bases $\left(\frac{\partial}{\partial x_i}\right), \left(\frac{\partial}{\partial y_j}\right)$, the matrix of the linear transformation $\Phi_*(X)$ is none other than the Jacobian matrix of the system (1.6).

More precisely, we have

$$e_i: f \to \left(\frac{\partial f^*}{\partial x_i}\right)_{\psi(p)} \quad (1 \leqslant i \leqslant m),\ f^* = f \circ \psi^{-1},$$

$$\bar{e}_j: g \to \left(\frac{\partial g^*}{\partial y_j}\right)_{\psi'(\Phi(p))} \quad (1 \leqslant j \leqslant n),\ g^* = g \circ (\psi')^{-1}$$

as basis vectors.

Then

$$\Phi_{*p}(e_i)g = e_i(g \circ \Phi) = \left(\frac{\partial (g \circ \Phi)^*}{\partial x_i}\right)_{\psi(p)}.$$

Since $(g \circ \Phi)^*(x_1, \ldots, x_m) = g^*(y_1, \ldots, y_n)$, where $y_j = \phi_j(x_1, \ldots, x_m)$, we have

$$\Phi_{*p}(e_i) = \sum_{j=1}^{n} \left(\frac{\partial \phi_j}{\partial x_i}\right)_{\psi(p)} \bar{e}_j$$

giving the Jacobian matrix.

The inverse function theorem of classical real-variable analysis now gives the result that if Φ_{*p} is an isomorphism of M_p onto $N_{\Phi(p)}$, then there exist open submanifolds $U \subset M$ and $V \subset N$ such that $p \in U$ and Φ is a diffeomorphism of U onto V.

The differentiable mapping $\Phi: M \to N$ is called **regular** at $p \in M$ if Φ_* is a bijective mapping of M_p into $N_{\Phi(p)}$.

1.11.1 Mappings and vector fields

Let $\Phi: M \to N$ be a differentiable map. Then, although Φ induces a linear map Φ_* which maps a vector at p to a vector at $\Phi(p)$, in general it does not map a vector field on M to a vector field on N. For example, it may happen that a different point $p' \in M$ will also map to $\Phi(p)$, and Φ_* will give two different image vectors at $\Phi(p)$. However, let X and Y be vector fields over M and N such that

$$\Phi_{*p}(X_p) = Y_{\Phi(p)}, \text{ for all } p \in M .$$

Then such vector fields are called **Φ-related**.

Exercise

Prove that if X_1 and Y_1 are Φ-related, and in addition X_2 and Y_2 are Φ-related, then $[X_1, X_2]$ and $[Y_1, Y_2]$ are Φ-related.

1.11.2 Mappings and differential forms

Let $\Phi: M \to N$ be a differentiable mapping. Then we show that Φ induces another linear-map Φ^* from the dual tangent space at $\Phi(p)$ to the dual tangent space at p. Let ω be a covariant vector at $\Phi(p)$, and let X_p be any tangent vector at p. We define the **pull-back** of ω by Φ as follows:

$$(\Phi^*\omega)_p X_p = \omega(\Phi_* X)_{\Phi(p)} .$$

Similarly an r-form $\omega_{\Phi(p)}$ can be pulled back to a r-form at p, by requiring

$$(\Phi^*\omega)_p(X_1, X_2, \ldots, X_r) = \omega_{\Phi(p)}(\Phi_* X_1, \ldots, \Phi_* X_r)$$

where X_i, $i = 1, \ldots, r$ are vectors at p. In fact, in this way we can pull back an r-form ω on N to an r-form on M. We merely require that if X_i, Y_i are Φ-related for $1 \leqslant i \leqslant r$, then

$$(\Phi^*\omega)(X_1, \ldots, X_r) = \omega(Y_1, \ldots, Y_r) .$$

We now compute $\Phi^*\omega$ in local coordinates. As before, denote by U, V open sets of M and N where local coordinates x_i, y_j are valid. Then an r-form ω has an expression

$$\omega = \Sigma g_{i_1} \ldots {}_{j_r} dy_{j_1} \wedge dy_{j_2} \wedge \ldots \wedge dy_{j_r}$$

where $g_{j_1} \ldots {}_{j_r} \in C^\infty(V)$.

The form $\Phi^*\omega$ induces on U an r-form with expression

$$\Phi^*\omega = \Sigma f_{i_1} \ldots i_r dx_{i_1} \wedge \ldots \wedge dx_{i_r} .$$

However, from the relation $y_j = \phi_j(x_1, \ldots, x_m)$ $\quad 1 \leqslant j \leqslant n$ we have

$$dy_j = \sum_{i=1}^{m} \frac{\partial \phi_j}{\partial x_i} dx_i$$

and hence

$$\Phi^*\omega = \Sigma g_{j_1} \ldots {}_{j_r} \frac{\partial \phi_{j_1}}{\partial x_{i_1}} \frac{\partial \phi_{j_2}}{\partial x_{i_2}} \cdots \frac{\partial \phi_{j_r}}{\partial x_{i_r}} dx_{i_1} \wedge dx_{i_2} \wedge \ldots \wedge dx_{i_r} .$$

From this coordinate representation it is easy to verify that if $\omega_1, \omega_2 \in \mathscr{A}(N)$, then

$$\Phi^*(\omega_1 \wedge \omega_2) = \Phi^*(\omega_1) \wedge \Phi^*(\omega_2) .$$

Moreover for any $\omega \in \mathscr{A}(N)$ we have

$$\Phi^* d\omega = d\Phi^*\omega ,$$

that is, the operations of exterior derivation and pull-back are commutative. These important properties will be much used in later chapters.

1.12 PARALLELISM

An affine connexion ∇ enables one to introduce the idea of parallelism between tangent vectors Y_p at p and Y_q at q, relative to a curve γ joining p to q. By a **curve** γ we mean a regular mapping of an open internal $I \subset \mathbb{R}$ into the manifold M. By a **curve segment** we mean a restriction of γ to a closed subinterval of I.

Let $\gamma: t \rightarrow \gamma(t)$ be a curve in M. Let $Y(t)$ be a tangent vector to M at each point of γ which varies differentiably with t. The tangent vector $X(t)$ to γ at t is the image of $\frac{d}{dt}$ under the differential of the map γ, that is, $X(t) = d\gamma\left(\frac{d}{dt}\right)_t$. Let J be a closed subinterval of I such that $\gamma(t)$ is contained in a coordinate neighbourhood U and has no double points in U. Then we would like to say that the vector field $Y(t)$ is **parallel along** $\gamma(t)$ if $\nabla_{X(t)} Y(t) = 0$. However, we have defined $\nabla_X Y$ only for vector fields defined over the open submanifold U. We have to show that there exist vector fields X, $Y \in \mathscr{D}^1(U)$ whose restriction to $\gamma(t)$ are $X(t)$ and $Y(t)$, so that the condition $\nabla_X Y = 0$ makes sense. Furthermore we have to show that the restriction of this condition to γ is independent of the particular extension fields considered—in particular, the condition depends only on the original fields $X(t)$, $Y(t)$, that is, the values of X and Y at points *on the curve*. Details of this argument will be found in Helgason (1978), p. 29. In terms of local coordinates x_i over U, writing

$$X = X^i \frac{\partial}{\partial x_i}, \quad Y = Y^i \frac{\partial}{\partial x_i} \text{ on } U \, ,$$

the condition for parallelism becomes

$$\frac{dY^k}{dt} + \Gamma^k_{ij} Y^i \frac{dx_j}{dt} = 0, \; t \in J \, .$$

The curve γ is called a **geodesic** if it is **self-parallel**, that is, if the tangent vector $d\gamma\left(\frac{d}{dt}\right)_t$ is parallel along the curve. The condition for this is

$$\frac{d^2 x^k}{dt^2} + \Gamma^k_{ij} \frac{dx_i}{dt} \frac{dx_j}{dt} = 0 \, .$$

1.13 RIEMANNIAN GEOMETRY

Let M be a C^∞-manifold. A **pseudo-Riemannian** structure on M is a tensor field g of type (0, 2) such that

(i) $g(X, Y) = g(Y, X)$ for $X, Y \in \mathscr{D}^1(M)$
(ii) for each $p \in M$, $g_p(X_p, Y_p) = 0$ for all Y_p implies $X_p = 0$.

An alternative way of expressing (ii) is to require that g_p is a non-degenerate bilinear form on $M_p \times M_p$.

If the requirement (ii) is strengthened so that g_p is positive definite for each p, then the structure is called **Riemannian**. We shall be concerned primarily with this case. The so-called **fundamental theorem of pseudo-Riemannian Geometry** states that *there exists one and only one affine connexion such that*:

(a) *the torsion tensor T is zero,*
(b) *parallel displacement along a curve preserves the inner product on tangent spaces.*

To prove the theorem we note that (a) implies

$$\nabla_X Y - \nabla_Y X = [X, Y], X, Y \in \mathscr{D}^1 .$$

Condition (b) implies

$$Zg(X, Y) = g(\nabla_X Z, Y) + g(X, \nabla_Z Y) + g([Z, X], Y) . \quad (1.7)$$

We permute the letters X, Y, Z cyclically to obtain two more relations like (1.7). We add two equations and subtract the third to get

$$2g(X, \nabla_Z Y) = Zg(X, Y) + g(Z, [X, Y]) + Yg(X, Z) + g(Y, [X, Z]) - Xg(Y, Z) - g(X, [Y, Z]) . \quad (1.8)$$

Since g is assumed non-degenerate, this shows that such a connexion must be unique. Conversely, if we take this as a definition for ∇, it is easily verified that conditions (a) and (b) are satisfied.

Writing

$$X = \frac{\partial}{\partial x_i}, Y = \frac{\partial}{\partial x_j}, Z = \frac{\partial}{\partial x_k}, g\left(\frac{\partial}{\partial x_i}, \frac{\partial}{\partial x_j}\right) = g_{ij} ,$$

we have

$$2g_{hi}\Gamma^h_{jk} = \partial_k g_{ij} + \partial_j g_{ik} - \partial_i g_{jk} ,$$

where we have written

$$\partial_k g_{ij} = \frac{\partial g_{ij}}{\partial x_k} .$$

Writing (g^{ij}) as the inverse matrix of (g_{ij}), which exists because of non-degeneracy, we get

$$\Gamma^i_{jk} = \tfrac{1}{2} g^{ih}(\partial_j g_{hk} + \partial_k g_{hj} - \partial_h g_{jk}) \ . \tag{1.9}$$

It is sometimes convenient to use symbols Γ_{jki} defined by $\Gamma_{jki} = g_{ih}\Gamma^h_{jk}$, and hence $\Gamma^h_{jk} = g^{hi}\Gamma_{jki}$. This formula gives the Christoffel symbols (connexion coefficients) for the required connexion, and it clearly shows that the connexion is uniquely determined by the tensor g. The results obtained in the section on affine connexions apply immediately to pseudo-Riemannian manifolds with this connexion. However, because of the special nature of this connexion, we find that the curvature tensor satisfies many relations which are no longer satisfied in the general case.

Theorem 4. *The curvature tensor of a pseudo-Riemannian manifold satisfies*

(i) $R(X, Y)Z + R(Y, X)Z = 0$,

(ii) $R(X, Y)Z + R(Y, Z)X + R(Z, X)Y = 0$ [1st Bianchi identity] ,

(iii) $< R(X, Y)Z, W > + < R(X, Y)W, Z > = 0$,

(iv) $< R(X, Y)Z, W > = < R(Z, W)X, Y >$.

Identity (i) follows immediately from the definition of R. To prove (ii), it will be sufficient to assume $[X, Y]$, $[Y, Z]$ $[X, Z]$ are all zero, as would happen if they were vectors of a coordinate basis. So we have to establish the identity

$$\begin{aligned} &- \nabla_X(\nabla_Y Z) + \nabla_Y(\nabla_X Z) \\ &- \nabla_Y(\nabla_Z X) + \nabla_Z(\nabla_Y X) \\ &- \nabla_Z(\nabla_X Y) + \nabla_X(\nabla_Z Y) = 0 \ . \end{aligned}$$

Because the connexion has zero torsion,

$$\nabla_Y Z - \nabla_Z Y - [Y, Z] = 0 \ ,$$

and hence the first and sixth terms of the above expression cancel. Similarly the second and third terms cancel, as do the fourth and the fifth. This establishes the result.

To prove (iii) it is sufficient to prove $< -\nabla_X\nabla_Y Z + \nabla_Y\nabla_X Z, Z > = 0$, that is,

$$< \nabla_X\nabla_Y Z, Z > = < \nabla_Y\nabla_X Z, Z > \ .$$

Since $[X, Y] = 0$, the expression $XY < Z, Z >$ is symmetric in X and Y.

Now $\quad X < Z, Z > = < \nabla_X Z, Z > + < Z, \nabla_X Z > = 2 < \nabla_X Z, Z >$.

Hence $\quad YX < Z, Z > = 2 < \nabla_Y\nabla_X Z, Z > + 2 < \nabla_X Z, \nabla_Y Z >$.

Thus the first term of the right-hand member is symmetric in X and Y, and we are home.

Exercise

Prove that (iv) follows from (i), (ii) and (iii). If you find this difficult, do not panic because we shall come across an alternative proof later.

A consequence of (iv) is that

$$< R(\partial_h, \partial_i)\partial_j, \partial_k > = R_{hijk} \ ;$$

the convention of Helgason (1978) and Kobayashi and Nomizu (1963) would have given a minus sign on the right.

1.14 NORMAL COORDINATES AND THE EXPONENTIAL MAP IN A RIEMANNIAN MANIFOLD

We have already seen that the curve $\gamma: I \to M$ is a geodesic if in terms of local coordinates it satisfies the differential equation

$$\frac{d^2 x_i}{dt^2} + \sum_{j,k=1}^{m} \Gamma^i_{jk} \frac{dx_j}{dt}\frac{dx_k}{dt} = 0 \ .$$

From standard theory of existence theorems of differential equations (for example, Graves, *Theory of Functions of Real Variables*, p. 166) we get

Theorem 5. *For every point p on a Riemannian manifold M there exists a neighbourhood U of p and a number $\varepsilon > 0$ such that, for each $q \in U$, and each tangent vector $v \in M_p$ with length $< \varepsilon$ there is a unique geodesic γ_v: $(-2, 2) \to M$ satisfying the conditions*

$$\gamma_v(0) = p, \quad \frac{d\gamma_v}{dt}(0) = v.$$

It follows that there exists a geodesic segment γ: $[0, 1] \to M$ satisfying the conditions $\gamma(0) = p$, $\dfrac{d\gamma}{dt}(0) = v$. The point $\gamma(1) \in M$ will be denoted by $\exp_p v$ and called the **exponential** of the tangent vector v. The geodesic segment can be described by the formula

$$\gamma(t) = \exp_p(tv) \ .$$

Let (e_i), $i = 1, \ldots, m$ be an orthonormal basis at p, and let u be the tangent vector of p which is mapped into a point q by the exponential map. Then, if $u = \Sigma x_i e_i$, the components x_i are called the **normal coordinates** of q.

The essential point of using normal coordinates is to simplify computations. We shall show that at the point p, the Christoffel symbols all vanish as do the first partial derivatives of the metric tensor. Certainly the equations for a geodesic must be satisfied by

$$x_i = a^i t.$$

Thus we have $\Gamma^i_{jk}(at)a^j a^k = 0$ for all values of a^i.

In particular at the point p we have

$$(\Gamma^i_{jk})_p a^j a^k = 0 \ ,$$

for all values of a^i from which it follows that $(\Gamma^i_{jk})_p = 0$.

Hence $\quad (\Gamma_{ijk})_p = (g_{kh})_p (\Gamma^h_{ij})_p = 0 \ ,$

and $\quad (\partial_j g_{ik})_p = (\Gamma_{ijk})_p + (\Gamma_{kji})_p = 0 \ .$

We have as a classical theorem:

Theorem 6. *A necessary and sufficient condition that* (x_i) *should be a normal coordinate-system of origin p is that*

$$g_{ij} x_j = (g_{ij})_p x_j \ .$$

For the proof of this see, for example, Ruse, Walker and Willmore (1961), page 12. As an application of normal coordinates we prove the second Bianchi identity, that is,

$$\nabla_i R_{jklp} + \nabla_j R_{kilp} + \nabla_k R_{ijlp} = 0 \ .$$

We have

$$\begin{aligned} R_{jklp} &= < R_{X_j X_k} X_l, X_p > \\ &= < \nabla_{[X_j, X_k]} X_l - [\nabla_{X_j}, \nabla_{X_k}] X_l, X_p > \\ &= (-\partial_j \Gamma^q_{lk} + \partial_k \Gamma^q_{lj} - \Gamma^h_{lk} \Gamma^q_{hj} + \Gamma^h_{lj} \Gamma^q_{hk}) g_{pq} \ . \end{aligned}$$

Taking covariant derivative with respect to x_i and evaluating at p this reduces to

$$\nabla_i R_{jklq} = \partial^2_{ik} \Gamma^q_{lj} - \partial^2_{ij} \Gamma^q_{lk} \ .$$

From this the second Bianchi identity follows immediately.

By using normal coordinates we may also prove the first Bianchi identity

$$R_{ijkl} + R_{iklj} + R_{iljk} = 0 \ .$$

Similarly we can prove identity (iv) of Theorem 4.

We note that by choosing normal coordinates at p, it follows that the connexion matrix (ω_{ij}) vanishes at this point.

1.15 THE LAPLACIAN

We denote by Δf the **Laplacian** of f, defined by

$$\Delta f = g^{ij}\nabla^2_{ij} f \ .$$

This is a natural generalization of the classical Laplacian in euclidean space, namely

$$\Delta f = \sum_{i=1}^{n} \left(\frac{\partial^2 f}{\partial x_i^2}\right) .$$

We write $g = \det(g_{ij})$. Then we have

$$\frac{\partial g}{\partial x^k} = g^{ij} g \partial_k g_{ij}$$

which gives

$$\frac{\partial}{\partial x^k}(\log g) = g^{ij}\partial_k g_{ij} \ .$$

Now

$$\begin{aligned}\Gamma^i_{ik} &= \tfrac{1}{2} g^{ih}\left[\partial_i g_{hk} + \partial_k g_{ih} - \partial_h g_{ik}\right] \\ &= \tfrac{1}{2} g^{ih}\partial_k g_{ih} \\ &= \tfrac{1}{2}\partial_k (\log g) \ .\end{aligned}$$

Let $X = X^i \dfrac{\partial}{\partial x^i}$ be a vector field defined over some coordinate neighbourhood U. Then the **divergence** of X is given by

$$\operatorname{div} X = \nabla_i X^i.$$

Clearly

$$\begin{aligned}\operatorname{div} X &= \partial_i X^i + \Gamma^i_{ji} X^j \\ &= \partial_i X^i + \tfrac{1}{2}\partial_j (\log g) X^j \\ &= \frac{1}{\sqrt{g}}\partial_i(\sqrt{g} X^i) \ .\end{aligned}$$

In particular when $X^i = g^{ij}\partial_j f$ we get

$$\Delta f = \frac{1}{\sqrt{g}}\partial_i(\sqrt{g} g^{ij}\partial_j f) \ .$$

1.16 IDENTITIES IN RIEMANNIAN GEOMETRY

We assume that we are working in normal coordinates, as this will simplify the form of some of the identities below.

We have already proved the Bianchi identities

$$R_{ijkl}+R_{iklj}+R_{iljk}=0\ , \tag{1.10}$$
$$\nabla_i R_{jklp}+\nabla_j R_{kilp}+\nabla_k R_{ijlp}=0\ . \tag{1.11}$$

In a similar way if we denote by R_{ij} the derivation of the tensor algebra determined by the curvature tensor, we have the Ricci identity

$$\nabla^2_{ji}-\nabla^2_{ji}=-R_{ij}\ . \tag{1.12}$$

We also have

$$\sum_i \nabla_i R_{iakl}=\nabla_k \rho_{al}-\nabla_l \rho_{ak} \tag{1.13}$$

where ρ_{al} are components of the Ricci tensor given by

$$\rho_{al}=\sum_b R_{babl}\ .$$

A!so

$$\sum_i \nabla_i \rho_{ij}=\tfrac{1}{2}\nabla_j \tau \tag{1.14}$$

where τ is the scalar curvature given by

$$\tau=\Sigma R_{ijij}\ .$$

As consequences of the first Bianchi identity (1.10) we have

$$\Sigma R_{abci}R_{acbj}=\tfrac{1}{2}\Sigma R_{abci}R_{abcj}\ , \tag{1.15}$$

$$\Sigma R_{abcd}R_{acbd}=\tfrac{1}{2}\Sigma R_{abcd}R_{abcd} \tag{1.16}$$

It is convenient to write

$$||\rho||^2=\Sigma\rho_{ij}\rho_{ij},\ ||R||^2=\Sigma R_{abcd}R_{abcd}\ .$$

Then (1.16) can be written

$$\Sigma R_{abcd}R_{acbd}=\tfrac{1}{2}||R||^2\ . \tag{1.16(a)}$$

It is convenient to introduce some new symbols as follows:

$$\check{\rho}=\Sigma\rho_{ij}\rho_{jk}\rho_{ki}\ ,$$
$$<\rho,\dot{R}>=\Sigma\ \rho_{ij}R_{ipqr}R_{jpqr}\ (\text{where } \dot{R}_{ij}=\Sigma\ R_{ipqr}R_{jpqr})\ ,$$
$$<\rho\otimes\rho,\bar{R}>=\Sigma\rho_{ij}\rho_{kl}R_{ikjl}\ (\text{where } \bar{R}_{ijkl}=R_{ikjl})\ ,$$
$$\check{R}=\Sigma\ R_{ijkl}R_{klpq}R_{pqij}\ ,$$
$$\check{\bar{R}}=\Sigma\ R_{ikjl}R_{kplq}R_{piqj}\ ,$$
$$||\nabla\tau||^2=\Sigma(\nabla_i\tau)^2\ ,$$
$$||\nabla\rho||^2=\Sigma(\nabla_i\rho_{jk})^2\ ,$$
$$\alpha(\rho)=\Sigma\ \nabla_i\rho_{jk}\nabla_k\rho_{ij}\ ,$$
$$||\nabla R||^2=\Sigma(\nabla_iR_{jklq})^2\ ,$$
$$<\Delta\rho,\rho>=\Sigma\ \rho_{ij}\nabla_{kk}\rho_{ij}\ ,$$
$$<\nabla^2\tau,\rho>=\Sigma\ (\nabla^2_{ij}\tau)\rho_{ij}\ ,$$
$$<\nabla R,R>=\Sigma\ R_{ijkl}\nabla^2_{pp}R_{ijkl}\ .$$

Then we have the following identities which we leave as an exercise to the reader to establish:

$$\Sigma\ \nabla_iR_{jabc}\nabla_iR_{jbac}=\tfrac{1}{2}||\nabla R||^2\ , \tag{1.17}$$
$$\Sigma\ \nabla_iR_{jabc}\nabla_jR_{iabc}=\tfrac{1}{2}||\nabla R||^2\ , \tag{1.18}$$
$$\Sigma\ \nabla_iR_{jabc}\nabla_jR_{ibac}=\tfrac{1}{4}||\nabla R||^2\ , \tag{1.19}$$
$$\Sigma\ R_{abcd}\nabla^2_{ii}R_{acbd}=\tfrac{1}{2}<R,\Delta R>\ , \tag{1.20}$$
$$\Sigma\ \rho_{ij}R_{iabc}R_{jbac}=\tfrac{1}{2}<\rho,\dot{R}>\ , \tag{1.21}$$
$$\Sigma\ R_{ijkl}R_{klpq}R_{piqj}=\tfrac{1}{2}\check{R}\ , \tag{1.22}$$
$$\Sigma\ R_{ijkl}R_{kplq}R_{piqj}=\tfrac{1}{4}\check{R}\ , \tag{1.23}$$
$$\Sigma\ R_{ijkl}R_{jplq}R_{pkqi}=\check{\bar{R}}-\tfrac{1}{4}\check{R}\ , \tag{1.24}$$
$$\Sigma\ (\nabla^2_{ij}\rho_{ik})\rho_{jk}=\tfrac{1}{2}<\nabla^2\tau,\rho>+\check{\rho}-<\rho\otimes\rho,\bar{R}>\ , \tag{1.25}$$
$$\Sigma\ (\nabla^2_{ij}R_{iabc})R_{jabc}=2\ \Sigma\ (\nabla^2_{ij}R_{iabc})R_{jbac}=\tfrac{1}{2}<\Delta R,R>\ , \tag{1.26}$$
$$\Sigma\ (\nabla^2_{ij}\rho_{kl})R_{ikjl}=<\nabla^2\rho,\bar{R}>=\tfrac{1}{4}<\Delta R,R>-\tfrac{1}{2}<\rho,\dot{R}> +\check{\bar{R}}+\tfrac{1}{4}\check{R}\ , \tag{1.27}$$
$$\Sigma\ \nabla^4_{ijij}\tau=\Sigma\ \nabla^4_{ijji}\tau=\Delta^2\tau+\tfrac{1}{2}||\nabla\tau||^2+<\nabla^2\tau,\rho>\ , \tag{1.28}$$
$$\Sigma\ \nabla^4_{ijki}\rho_{jk}=\Sigma\ \nabla^4_{ijkj}\rho_{ik}=\tfrac{1}{2}\Delta^2\tau+\tfrac{1}{2}||\nabla\tau||^2-2||\nabla\rho||^2 +2<\nabla^2\tau,\rho>-<\Delta\rho,\rho>+3\alpha(\rho)+2\check{\rho}-2<\rho\otimes\rho,\bar{R}> -\tfrac{1}{4}<\Delta R,R>+\tfrac{1}{2}<\rho,\dot{R}>-\check{\bar{R}}-\tfrac{1}{4}\check{R}\ , \tag{1.29}$$

$$\begin{aligned}\Sigma\, \nabla^4_{ijkk}\rho_{ij} = {} & \tfrac{1}{2}\Delta^2\tau + \tfrac{1}{2}||\nabla\tau||^2 + 4\alpha(\rho) + 2 < \nabla^2\tau, \rho > \\ & - 3||\nabla\rho||^2 - <\Delta\rho, \rho> + 2\check{\rho} - 2 < \rho \otimes \rho, \bar{R} > \\ & - \tfrac{1}{2} <\Delta R, R> + <\rho, \dot{R}> - 2\check{\bar{R}} - \tfrac{1}{2}\check{R} \,. \end{aligned} \tag{1.30}$$

To prove (1.17)–(1.24) one makes repeated use of (1.10) and (1.11). The Ricci identity (1.12) is used together with (1.10), (1.11) to prove the rest of the equations.

1.17 SECTIONAL, RICCI AND SCALAR CURVATURES

From the curvature tensor, we can construct three quantities all of which have geometric significance. Given a point p on the manifold and a 2-dimensional subspace of the tangent space at p, we can form the **sectional curvature** of the manifold at this plane. More specifically, let the subspace be determined by unit vectors X, Y at p. Then the sectional curvature is given by $R(X, Y, X, Y)$. The symmetry and skewsymmetry properties of R imply that $R(X, Y, X, Y)$ depends only on the 2-dimensional subspace generated by X, Y and is independent of the particular pair of unit vectors chosen to generate this subspace.

Given a point and a unit tangent vector X at that point, we can form the **Ricci curvature** in this direction by averaging all the sectional curvatures of the two dimensional tangent planes which contain this tangent. This is given relative to an orthonormal basis (e_i) by

$$\begin{aligned}\rho(X) &= \sum_{i=1}^{n} R(e_i, X, e_i, X) \\ &= \sum_{i,j=1}^{n} \rho_{ij} X^i X^j \,,\end{aligned}$$

where (ρ_{ij}) are components of the Ricci tensor and

$$X = \sum_{i=1}^{n} X^i e_i \,.$$

Given a point, we can form the **scalar curvature** at this point by averaging all the sectional curvatures at this point. We have

$$\tau = \sum_{i,j=1}^{n} R(e_i, e_j, e_i, e_j) \,.$$

Clearly a knowledge of the sectional curvature gives more information than a knowledge of either the Ricci curvature or the scalar curvature. The problem of finding conditions for the global existence on differentiable manifolds of Riemannian metrics with certain curvature conditions is surprisingly difficult. In the next chapter we shall show that the Gauss-

Bonnet theorem gives a satisfactory solution to this problem for 2-dimensional compact surfaces. In higher dimensions the problem becomes more complicated, although in even dimensions the generalized Gauss-Bonnet theorem does give some information.

Very recent results, obtained using methods too advanced to be summarized here, show that every compact manifold with dimension greater than two admits a Riemannian metric with negative scalar curvature. Moreover, it is known that every noncompact manifold admits a complete metric with negative scalar curvature. Thus, in higher dimensions, the existence of complete metrics with negative scalar curvature imposes no topological restrictions on the manifold. However, complete metrics with non-negative scalar curvature do give topological restrictions. One of the outstanding unsolved problems is to obtain a criterion for a manifold to admit a metric with positive scalar curvature.

The corresponding problems for Ricci curvature are even more difficult. A recent result shows that a compact manifold with non-negative Ricci curvature must have a finite fundamental group. Indeed, it seems probable that a complete manifold with positive Ricci curvature has a finite fundamental group, but this remains an open problem. One problem which attracted attention for many years was whether there exists a non-flat compact Riemannian manifold with zero Ricci curvature. We now know that such manifolds do in fact exist. These problems have been solved by methods involving differential geometry, algebraic geometry, topology and partial differential equations, thus illustrating the essential unity of modern mathematics.

1.18 ORIENTABLE MANIFOLDS

It is clear that the space of alternating tensors of order n based on an n-dimensional vector space is 1-dimensional. The differentiable manifold M is called orientable if it admits a continuous exterior differential form of degree n which is nowhere zero. Two such forms define the same orientation if they differ from one another by a factor which is everywhere positive.

An orientable manifold has exactly two possible orientations. Let ω, ω' be two n-forms which determine an orientation of M. Then $\omega' = f\omega$ and the function f is either everywhere positive or everywhere negative. Thus the only possible orientations are given by ω and by $-\omega$. The manifold M is called oriented if such a form ω is given, and $\omega \neq 0$ at any point m. The restriction of the form to a point m of M determines an orientation in each tangent space of M. Any differential form of degree n can then be written as $f(m)\omega$: we shall say that the form >0, <0, or $=0$ at m according as $f(m)>0$, <0 or $=0$.

The **support** of a scalar function f in M is the closure of the set of points of M where $f \neq 0$. Its complement is the maximal open set where $f=0$. More generally, the **support of a form** θ is the closure of the set of points of M where $\theta \neq 0$.

An open covering of M is said to be **locally finite** if any compact subset of M meets only a finite number of its members.

Theorem 7. *Let K be a family of open sets of M which form a base for the topology of M. Then there is a locally finite open covering of M whose members are sets of K.*

We make use of the following lemma from the topology of Hausdorff spaces.

Lemma *Any Hausdorff space possesses a countable covering $\{P_i\}$ of open sets P_i with compact closures.*

From the lemma it follows that M has this property. Write

$$Q_j = \bigcup_{1 \leqslant i \leqslant j} \bar{P}_i \ .$$

Then $\{Q_j\}$ form a countable covering of M by compact sets Q_j such that $Q_j \subset Q_{j+1}$. We now construct compact subsets R_j such that

$$Q_j \subset R_j, \ R_j \subset \text{Interior of } R_{j+1} \ .$$

We use the method of induction. Suppose that $R_1, R_2, \ldots, R_j$ have been constructed. Since $R_j \cup Q_{j+1}$ is compact, it has a finite covering by open sets with compact closures. We define R_{j+1} to be the union of these closures.

Let $S_j =$ Interior of R_j: let

$$T_j = R_j \cap (M \sim S_{j-1}) \ ,$$

where we agree that sets with negative indices are void. Since $R_{j-2} \subset S_{j-1}$ and $T_j \subset (M \sim S_j)$ we have

$$T_j \cap R_{j-2} = \varnothing \ .$$

For $m \in T_j$, there is a set of K containing m, contained in S_{j+1} and not meeting R_{j-2}. These sets, for all $m \in T_j$, form a covering of T_j. On the other hand, T_j is a closed subset of a compact set R_j, and hence is compact. Thus the above covering has a finite subcovering, which we call K_j. We denote by K' the family of the sets of K_j for all j.

The sets of K' form a covering of M. Indeed, if $m \in M$, there is a $k > 0$ such that $m \in R_k$, $m \notin R_{k-1}$. Hence $m \in T_k$ and is covered by a set of K'. Moreover, if $j \geqslant k+2$, R_k meets no set of K_j. Since every compact set of M is contained in a certain R_k, it follows that the covering K' is locally finite. This completes the proof of the theorem.

1.19 PARTITIONS OF UNITY

We are now in a position to return to partitions of unity already mentioned in a previous section.

Theorem 8. *Let $\{U_i\}$ be an open covering of a manifold M of class C^∞. Then there are functions g_α which satisfy the conditions:*

(1) *Each g_α is C^∞ and satisfies the inequality $0 \leqslant g_\alpha \leqslant 1$. Its carrier is compact and contained in one of the U_i.*
(2) *Every point of M has a neighbourhood which meets only a finite number of the carriers of g_α.*
(3) $\Sigma g_\alpha = 1$.

To prove the theorem we make use of the following lemma, similar to that of section 1.3:

Lemma *Let V be an open set in M with compact closure $\bar{V}$, which is contained in some coordinate neighbourhood U. Then there exists a real-valued function h defined on M, of class C^∞, such that $0 \leqslant h(m) \leqslant 1$ for all $m \in M$, $h(m) = 1$ for $m \in \bar{V}$, and $h(m) = 0$ for $m \in (M \smallsetminus U)$.*

We now consider open sets W with the following properties:

(1) The closure $\bar{W}$ is compact and contained in a certain U_i.
(2) There is a real-valued function w of class C^∞ on M, satisfying the inequalities $0 \leqslant w \leqslant 1$, such that W is the set of all points at which $w \neq 0$.

From the above lemma it follows that the family of all such sets W constitutes a base of M. From Theorem (8) it follows that there is a locally finite orientable open covering of M formed by the sets W. Denote such a covering by $\{W_\alpha\}$, and denote the function corresponding to W_α by w_α. Since the covering is locally finite, every point $m \in M$ has a neighbourhood which meets only a finite number of W_α, that is, in which only a finite number of the functions w_α take non-zero values. *The sum* $\sum_\alpha w_\alpha(m)$ *is therefore a finite sum* and is different from zero. We define

$$g_\alpha(m) = \frac{w_\alpha(m)}{\sum_\alpha w_\alpha(m)} .$$

Then it is easily checked that these functions satisfy the conditions of the theorem.

1.20 APPLICATION OF PARTITIONS OF UNITY

1.20.1 Existence of Riemannian metrics on *M*

There appear to be essentially three ways of proving that a C^∞-differentiable manifold always supports a C^∞-Riemannian metric. One way is to make use of Whitney's imbedding theorem which states that any differentiable manifold of dimension n can be imbedded in a euclidean space of dimension $2n+1$. Such an imbedding induces a Riemannian metric on the manifold from the euclidean metric of the containing space. Whitney's theorem is, however, quite difficult to prove.

A second way is to use the conditions for the existence of a global cross-section of the fibre-bundle of metrics (cf. Steenrod, 1951, page 58).

We use a third method in order to illustrate the use of partitions of unity. Suppose we have a locally finite covering $\{U_i\}$ of M by coordinate neighbourhoods. For each U_i, consider some positive definite quadratic form ds_i^2 defined over it—for example, we could take the flat euclidean metric. Let g_α be a corresponding partition of unity. Consider the quadratic differential form

$$\Sigma\, g_\alpha ds_i^2 \ ,$$

when i is an index such that U_i contains the support of g_α. This clearly defines a Riemannian metric over M.

1.20.2 Application to Integration over Manifolds

This subject is non-trivial. There are essentially two different ways of defining integration of forms over a differentiable manifold—one by considering the manifold as a simplicial complex as in algebraic topology, and the second by means of partition of unity. We adopt the second method. We have the fundamental existence theorem:

Theorem 9. *Let M be an oriented manifold of dimension n. Then there is one, and only one, functional which assigns to a continuous differential form Φ of degree n with compact support, a real number called the integral of Φ over M, denoted by $\int\Phi$ such that*

(*i*) $\int(\Phi_1+\Phi_2)=\int\Phi_1+\int\Phi_2$.

(*ii*) *If the support of Φ is contained in a coordinate neighbourhood U with coordinates $u_1, \ldots, u_n$, such that $du_1\wedge\ldots\wedge du_n>0$, and if, in terms of these coordinates Φ has expression*

$$\Phi(u_1, \ldots, u_n)\, du_1\wedge\ldots\wedge du_n \ ,$$

then $$\int\Phi=\int_U\Phi(u_1, \ldots, u_n)\, du_1\wedge\ldots\wedge du_n$$

where the last member is a Riemann integral.

Let Φ be given over M with compact support S. Let $\{U_i\}$ be an open covering of M by coordinate neighbourhoods and let g_α be a corresponding partition of unity. Then every point $m\in S$ has a neighbourhood V_m which meets only a finite number of the supports of g_α. These V_m for all $m\in S$ form a covering of S. Since S is compact, it has a finite sub-covering. Thus there are only a *finite* number of $g_\alpha\Phi$ which are not identically zero.

We define

$$\int\Phi=\sum_\alpha \int g_\alpha\Phi \ ,$$

where the right-hand side is a finite sum. Since the differential form in each summand has a support which lies in a coordinate neighbourhood U_i, it can be evaluated according to the formula in condition (ii).

We now have to justify the above definition—in particular we have to show that it is independent of the choice of the neighbourhood U_i which contains the support $g_\alpha\Phi$, and in addition we have to show that it is independent of the choice of U_i and its associated partition of unity g_α.

To prove the first part is easy. For suppose that the support of $g_\alpha\Phi$ lies in two coordinate neighbourhoods U_i, U_j with local coordinates $(u_1, \ldots, u_n)$, $(v_1, \ldots, v_n)$ respectively. We take an open set $W \subset U_i \cap U_j$ which contains the support. In W we have

$$\begin{aligned} g_\alpha\Phi &= \Phi(u_1, \ldots, u_n)\, du_1 \wedge \ldots \wedge du_n \\ &= \Psi(v_1, \ldots, v_n)\, dv_1 \wedge \ldots \wedge dv_n \end{aligned}$$

where $\Psi(v_1, \ldots, v_m) = \Phi(u_1(v_1, \ldots, v_m), \ldots, u_n(v_1, \ldots, v_m)) \dfrac{\partial(u_1, \ldots, u_n)}{\partial(v_1, \ldots, v_n)}$,

the Jacobian determinant being positive throughout W. Then the equation

$$\int_W \Phi\, (du_1 \wedge \ldots \wedge du_n) = \int_W \Phi \frac{\partial(u_1, \ldots, u_n)}{\partial(v_1, \ldots, v_m)}\, dv_1 \wedge \ldots \wedge dv_n$$

is just the classical formula for the transformation of multiple integrals. It follows that our definition is independent of the particular choice of the neighbourhood U_i.

To prove the independence of the choice of cover $\{U_i\}$ with functions g_α, consider a second covering $\{V_j\}$ with corresponding functions g'_β. Then $\{U_i \cap V_j\}$ will be a covering of M, with functions $g_\alpha g'_\beta$ as a corresponding partition of unity.

Then it follows that

$$\sum_\alpha \int g_\alpha\Phi = \sum_{\alpha,\beta} \int g_\alpha g'_\beta\Phi\,,$$

and

$$\sum_\beta \int g'_\beta\Phi = \sum_{\alpha,\beta} \int g_\alpha g'_\beta\Phi\,,$$

thus proving the independence.

In many of our applications we can use a partition of unity argument to reduce the practical problem of evaluating the integral to one where the support of the corresponding form lies inside a coordinate neighbourhood, and the problem reduces to the evaluation of a (classical) multiple integral.

1.21 SUBMANIFOLDS

We have already defined what we mean by an open submanifold of the differentiable manifold M. We now define the more general concept of submanifold which includes open submanifold as a particular case.

A **submanifold** of a differentiable manifold M is a pair (N, ϕ) where ϕ is a differentiable map of a differentiable manifold N into M such that

(i) ϕ is injective,

(ii) for every $q \in N$, ϕ_* is injective—alternatively we require that ϕ_* has maximal rank.

We see that our previous definition of an open submanifold is consistent with the above, by taking ϕ as the inclusion map.

A **closed submanifold** is a submanifold (N, ϕ) such that

(i) $\phi(N)$ is a closed subset of M ,

(ii) every point $p \in N$ is contained in a coordinate neighbourhood U with local coordinates $x_1, \ldots, x_m$ such that the set $\phi(N) \cap U$ is defined by the equations

$$x_{r+1}=0, x_{r+2}=0, \ldots, x_m=0 .$$

The following example shows that although ϕ is differentiable and injective, the inverse map ϕ^{-1} need not be continuous, that is, ϕ is not necessarily an injective homeomorphism. Let M be the torus defined by coordinates (x, y) where x, y are real numbers modulo 1. Let N be the real line R^1, $-\infty < t < \infty$, and let ϕ be the map $t \to (t, \sqrt{2}t)$. Then (N, ϕ) is a submanifold of M. However the image $\phi(N)$ is everywhere dense in M, so ϕ cannot be a homeomorphism.

We have defined our submanifold as an **imbedding** because we have insisted that ϕ is injective. Often it is convenient to consider a more general map called an **immersion**. We obtain this by relaxing the requirement that ϕ should be injective, but replacing this by imposing the condition that if $\phi(q)=\phi(q')$ for $q' \neq q$, then $\phi_*(N_q)$ and $\phi_*(N_{q'})$ have just the zero vector in common. Thus a curve on M is allowed to self-intersect provided that the tangent lines at the point of intersection are distinct.

Returning to our previous example, we have seen that ϕ need not be an injective homeomorphism. We say that (N, ϕ) is **regularly imbedded** in M if ϕ is indeed an injective homeomorphism. A standard theorem states that *a submanifold* (N, ϕ) *is regularly imbedded if and only if it is a closed submanifold of* M. We shall only need that part of the theorem which asserts that a closed submanifold is regularly imbedded and this is clearly true.

We are now in a position to prove Stoke's formula which may be regarded as the fundamental theorem of the exterior differential calculus, analogous to the fundamental theorem of classical calculus. To describe the formula we must first clarify what we mean by a manifold with a boundary.

1.22 STOKE'S FORMULA

This formula gives a relation between an integral over a domain and another integral taken over its boundary. It includes as a special case Green's Theorem and Stoke's Theorem of classical analysis.

Let M be a differentiable manifold of dimension n. A **domain D with regular boundary** is a subset of M such that if $p \in D$, either (i) p has a neighbourhood belonging entirely to D, or (ii) there is a coordinate neighbourhood about p with coordinates $u_1, \ldots, u_n$ such that $U \cap D$ is the set of all points $q \in D$ with $u_n(q) \geqslant u_n(p)$.

Clearly a point p possesses either property (i) or property (ii) but not both. Points with property (i) are called interior points of D, and those with property (ii) are called boundary points. The set of all such boundary points is called the **boundary** of D, and is denoted by ∂D. We now have

Theorem 10. *The boundary of a domain with regular boundary is a closed submanifold which is regularly imbedded. If M is orientable so is the boundary* ∂D.

The first part follows from the previous theorem of Section 1.21 because the set $\partial D = \bar{D} \smallsetminus D$ and is closed.

To prove the second part, suppose that M is oriented. Let $p \in \partial D$. We choose a coordinate system around p such that

$$du_1 \wedge \ldots \wedge du_n > 0 \ .$$

Then we assert that ∂D can be oriented by requiring that

$$(-1)^n du_1 \wedge \ldots \wedge du_{n-1} > 0 \text{ at } p \ .$$

We have to prove that this condition is independent of the particular coordinate system chosen. To do this, let (v_i) be another coordinate system having property (ii). Since points on ∂D satisfy $u_n(q) = u_n(p)$, $v_n(q) = v_n(p)$ where q is a point in a neighbourhood of p, it follows that in the coordinate transformation from u_n to v_n, that v_n is a function of u_n only. Thus we have at p

$$\frac{\partial(v_1, \ldots, v_n)}{\partial(u_1, \ldots, u_n)} = \frac{\partial(v_1, \ldots, v_{n-1})}{\partial(u_1, \ldots, u_{n-1})} \frac{\partial v_n}{\partial u_n} > 0$$

since $$\frac{\partial v_n}{\partial u_n} > 0 \ .$$

Thus $(-1)^n dv_1 \wedge \ldots \wedge dv_{n-1} > 0$ on ∂D at p, thereby showing that our criterion for orientability of ∂D is independent of the coordinate system chosen.

At every point on the $(n-1)$-dimensional manifold ∂D we have a definite sign of a non-zero form of degree $(n-1)$. The bundle of $(n-1)$ forms over ∂D has fibres homeomorphic to the euclidean line $\mathbb{R}^1$. If we delete the zero form, then the fibres will be disconnected. However, by defining the sign of the non-zero form at each point of ∂D, we can choose at each point of ∂D one of the two components and we can do this in a continuous manner over ∂D. Thus we obtain the bundle of positive forms over ∂D, with fibre homeomorphic to the euclidean line $\mathbb{R}^1$. From the previous section about fibre bundles, this property ensures the existence of the global cross-section. Thus ∂D becomes an oriented submanifold.

We now consider a domain D with regular boundary ∂D and suppose it is compact. Define the characteristic function $h(p)$, $p \in M$ by

$$h(p)=1,\ p\in D;\ h(p)=0,\ p\in M \smallsetminus D.$$

We define the integral over D of an n-dimensional form Φ on M by

$$\int_D \Phi = \int h\Phi\ .$$

1.22.1 Stoke's Theorem

Let ω be a form of degree $(n-1)$ in M. Then

$$\int_D d\omega = \int_{\partial D} \omega\ .$$

To prove the formula let $\{U_i\}$ be an open covering of M by coordinate neighbourhoods such that for each U_i either $U_i \cap \partial D = \varnothing$ or U_i has property (ii). Let g_α be a corresponding partition of unity. Since ∂D and D are both compact, each meets only a finite number of the supports of g_α and hence

$$\int_{\partial D} \omega = \sum_\alpha \int_{\partial D} g_\alpha \omega\ ,$$

$$\int_D d\omega = \sum_\alpha \int_D d(g_\alpha \omega)\ .$$

Thus it is sufficient to establish Stoke's formula when the support of ω lies in a single coordinate neighbourhood U_i. We show that this leads to a consideration of only three possible cases and we establish the formula separately in each case. Let $u_1, \ldots, u_n$ be local coordinates in U_i such that

$$du_1 \wedge \ldots \wedge du_n > 0$$

and let

$$\omega = \sum_{j=1}^{n} (-1)^{j-1} a_j du_1 \wedge \ldots \wedge du_{j-1} \wedge du_{j+1} \wedge \ldots \wedge du_n\ .$$

Then we have

$$d\omega = \left(\sum_{j=1}^{n} \frac{\partial a_j}{\partial u_j}\right) du_1 \wedge \ldots \wedge du_n \ .$$

Now we first consider the case when $U_i \cap \partial D = \varnothing$. Then $\int_{\partial D} \omega = 0$. Now there are two possibilities—either $U_i \subset M \smallsetminus D$ or U_i lies in the interior of D. If the first possibility holds, then the left-hand side of the formula is zero and we are home. Now suppose that U_i lies in the interior of D, so that the left-hand side of the formula is equal to

$$\int_D d\omega = \int_C \left(\sum_{j=1}^{n} \frac{\partial a_j}{\partial u_j}\right) du_1 du_2 \ldots du_n$$

where C is a cube in the space of coordinates which contains the support of ω in its interior. By taking m sufficiently large we can suppose that C is defined by the inequalities $|u_j| < m$. The integral is now just an iterated integral in the sense of classical analysis. Thus we have

$$\int_C \frac{\partial a_j}{\partial u_j} du_1, \ldots, du_n = \pm \int_{F_j} \{a_j(u_1, \ldots, u_{j-1}, m, u_{j+1}, \ldots, u_n)$$
$$- a_j(u_1, \ldots, u_{j-1}, -m, u_{j+1}, \ldots, u_n)\} du_1 \ldots du_{j-1} du_{j+1} \ldots du_n \ ,$$

where F_j is the union of the appropriate faces of the cube. However, since the support of ω lies inside C, the above integrand vanishes. So, again the left-hand side of the formula is zero, and we are home.

So we are left with the third possibility that U_i is a coordinate neighbourhood with property (ii). We assume that ∂D is contained in the subset given by $u_n = 0$. When $u_n(p) \geqslant 0$ the corresponding characteristic function $h(p) = 1$. Now take in the space of coordinates u_j a cube C defined by

$$|u_k| \leqslant m, \ k = 1, \ldots, n-1; \ m \geqslant u_n > 0 \ ,$$

such that the support of ω is contained in the union of its interior and the side $u_n = 0$. As in the above analysis we have

$$\int_C \frac{\partial a_k}{\partial u_k} du_1 \ldots du_n = 0, \text{ for } k = 1, 2, \ldots, n-1 \ .$$

Also

$$\int_C \frac{\partial a_n}{\partial u_n} du_1 \ldots du_n = (-1)^{n-1} \int_{\partial D} a_n(u_1, \ldots, u_{n-1}, 0) \, du_1 \ldots du_{n-1} \ .$$

But the right-hand side of the last equation is just

$$\int_{\partial D} \omega \ .$$

Thus we have established the formula of Stokes.

It should be stated that the above treatment of integration on manifolds and Stoke's formula follows very closely that given by S. S. Chern in a series of lecture notes given at the University of Chicago 1959. A very similar treatment will be found in Matsushima (1972). We note that if the manfold M is given as compact, our treatment would be simplified as we could then avoid introducing partitions of unity.

1.23 EXERCISES

(1) Suppose $K \subset \mathbb{R}^2$ is a compact set with boundary ∂K, so that ∂K is a curve in $\mathbb{R}^2$. Let α be the 1-form given by $\alpha = P(x, y)dx + Q(x, y)dy$ where P, Q are differentiable functions defined in some open set $U \supset K$. Prove that

$$\iint_K \left(\frac{\partial Q}{\partial x} - \frac{\partial P}{\partial y}\right) dx \wedge dy = \int_{\partial K} Pdx + Qdy \ .$$

(2) Denote the coordinates of $\mathbb{R}^3$ by (x, y, z). Let M be a surface in $\mathbb{R}^3$, supposed oriented, and let $K \subset M$ be a compact set with a boundary. Let $\alpha = Pdx + Qdy + Rdz$ be a 1-form with P, Q, R differentiable functions of x, y, z in a neighbourhood of K. Prove that

$$\iint_K \left(\frac{\partial R}{\partial y} - \frac{\partial Q}{\partial z}\right) dy \wedge dz + \left(\frac{\partial P}{dz} - \frac{\partial R}{\partial x}\right) dz \wedge dx + \left(\frac{\partial Q}{\partial x} - \frac{\partial R}{\partial y}\right) dx \wedge dy$$
$$= \int_{\partial K} Pdx + Qdy + Rdz \ .$$

(3) Let K be a compact subset of $\mathbb{R}^3$ with a boundary ∂K, so that ∂K is a surface in $\mathbb{R}^3$. Let $\alpha = Ady \wedge dz + Bdz \wedge dx + Cdx \wedge dy$ be a 2-form, where A, B, C are differentiable functions of x, y, z in a neighbourhood K. Prove that

$$\iiint_K \left(\frac{\partial A}{\partial x} + \frac{\partial B}{\partial y} + \frac{\partial C}{\partial z}\right) dx \wedge dy \wedge dx = \iint_{\partial K} Ady \wedge dz + Bdz \wedge dx + Cdx \wedge dy$$

(4) Let M be a Riemannian manifold. We define the "flip map" at a point $m \in M$. We take a point p in some normal coordinate neighbourhood based at m, and let p' be the point on the unique geodesic from p to m continued beyond m so that p, p' are equidistant from m. The map s_m which sends p into p' is called the flip map at m. Clearly, if r is the distance of p from m, s_m maps the geodesic sphere centre m radius r onto itself by mapping points into antipodal points. M is said to have a D'Atri metric if for each m, the maps s_m are volume preserving, that is, the

volume element at p is equal to the volume element at p'. An interesting (but at present unsolved) problem is to classify such metrics. Prove that

$$ds^2 = dx^2 + dz^2 + (z\,dx - dy)^2$$

is a D'Atri metric on $\mathbb{R}^3$.

Chapter 2

The Gauss-Bonnet formula

2.1 INTRODUCTION

In this chapter we shall apply the tools described in chapter 1 to prove one of the most fundamental results in global Riemannian geometry, namely, the Gauss-Bonnet formula. This expresses the Euler characteristic of the manifold in terms of an integral involving the curvature of the Riemannian structure. Thus the differential geometry and the topology of the manifold are related by the formula. The classical Gauss-Bonnet theorem for closed 2-dimensional surfaces is as follows:

Theorem 11. *Let M be a closed orientable Riemannian manifold of dimension 2. Then the total Gaussian curvature of M given by*

$$\int_M K dS$$

is equal to $2\pi\chi(M)$, *where* $\chi(M)$ *is the Euler characteristic of M.*

We shall prove the theorem using a triangulation of the surface involving f triangles, with a total of v vertices and e edges, in terms of which the Euler characteristic is given by the well-known formula

$$\chi(M) = v - e + f \,. \tag{2.1}$$

Then we shall prove as a corollary the following:

Theorem 12. *On a closed orientable surface, the sum of the indices of a vector field with a finite number of singularities, is equal to the Euler characteristic* $\chi(M)$ *of M.*

Instead of obtaining theorem 12 as a corollary to theorem 11 we give an alternative proof, due to L. Woodward, which is direct. We can then obtain theorem 11 as a consequence of theorem 12. We do this because in the corresponding Gauss-Bonnet theorem for a $2n$-dimensional Riemannian manifold, it is convenient to define the Euler characteristic in terms of the sum of indices of a vector field, and the $2n$-dimensional proof follows closely the ideas of the 2-dimensional proof.

2.2 THE THEOREM OF TURNING TANGENTS

We state and prove this theorem first for a closed curve in the euclidean plane E^2 but we shall see that, by using isothermal parameters, the same theorem and proof apply to closed curves in 2-dimensional surfaces in general. Our treatment of this theorem follows closely that of S. S. Chern, *Studies in Global Geometry and Analysis*, Math. Assoc. of America, 1967, p. 16.

We assume that E^2 is oriented with a prescribed sense of rotation. Let $\gamma : s \to \gamma(s)$ be a smooth curve in E^2 with arc length s. We assume that the curve consists only of regular points, that is, the tangent vector $\gamma'(s)$ is nowhere zero. We denote by $e_1(s)$ the unit tangent vector in the direction of $\gamma'(s)$, and by $e_2(s)$ the unit normal vector such that the rotation from e_1 to e_2 is positive. Then the position vector $\gamma(s)$ and the vectors $e_1(s)$, $e_2(s)$ are related by the Frenet formulae

$$\frac{d\gamma}{ds}(s) = e_1\ , \quad \frac{de_1}{ds} = \kappa e_2\ , \quad \frac{de_2}{ds} = -\kappa e_1\ ,$$

where κ is the curvature of the curve.

We shall assume that γ is closed and of length L so that $\gamma(s)$ is a periodic function of s of period L. Also we shall assume that the curve is simple, that is, does not self-intersect. More precisely, we assume that $\gamma(s_1) \neq \gamma(s_2)$ whenever $0 < s_1 - s_2 < L$. Our first task is to define what we mean by the **rotation index** of γ. Let O be a fixed point in E^2 which we regard as the origin of an orthogonal cartesian system of coordinates. Let Γ denote the unit circle, centre O. We define the **tangential mapping** $T : \gamma \to \Gamma$ as one which maps a point P of γ to the end point of the unit vector through O which is parallel to the tangent vector to γ at P. It is intuitively clear that T is continuous, and that as P goes round γ once, the image $\Gamma(P)$ will traverse Γ an integral number of times. This number is called the **rotation index** of γ. The theorem of turning tangents asserts that if γ is simple, then the rotation index is ± 1. This may seem so obvious as to require no proof—but, like the Jordan curve theorem which is even more obvious, a rigorous proof is non-trivial.

We take a fixed vector through O, say Ox, and denote by $\tau(s)$ the angle which $e_1(s)$ makes with Ox. We agree that $0 \leqslant \tau(s) < 2\pi$ so that $\tau(s)$ is uniquely determined. However the function $\tau(s)$ is not continuous. For that reason we construct a function $\tilde{\tau}(s)$, closely related to $\tau(s)$, which nevertheless is continuous. In fact, we prove that there exists a continuous function $\tilde{\tau}(s)$ such that $\tilde{\tau}(s) = \tau(s)$, mod 2π. Since T is a continuous function on a bounded closed interval, it is uniformly continuous and hence we can find a number δ such that for, $|s_1 - s_2| < \delta$, $T(s_1)$ and $T(s_2)$ lie in the same open half-plane. Hence it follows that if $\tilde{\tau}(s_1)$ is known, $\tilde{\tau}(s_2)$ is determined. We divide the interval $0 \leqslant s \leqslant L$ into subintervals by points $s_0 = 0 < s_1 < s_2 \ldots < s_m = L$ such that $|s_i - s_{i-1}| < \delta$, $i = 1, 2, \ldots, m$. We assign to s_0 the value $\tilde{\tau}(s_0) = \tau(s_0)$. Then $\tilde{\tau}(s)$ is determined in the interval $s_0 \leqslant s \leqslant s_1$ and hence at s_1. This in turn determines $\tilde{\tau}(s)$ in the second interval and hence at s_2.

Proceeding in this way, the function $\tilde{\tau}(s)$ is determined over the whole curve γ since only a finite number of operations are involved. The difference $\tau(L)-\tau(0)$ is an integral multiple of 2π, say $r(\gamma)2\pi$. We now show that $r(\gamma)$ is independent of the choice of the function $\tilde{\tau}(s)$. For if $\tilde{\tau}'(s)$ is another function satisfying our requirements, then

$$\tilde{\tau}'(s)-\tilde{\tau}(s)=n(s).2\pi$$

where $n(s)$ is an integer. However since $n(s)$ is continuous it must be a constant. Thus $\tilde{\tau}'(L)-\tilde{\tau}'(0)=\tilde{\tau}(L)-\tilde{\tau}(0)$ showing that $r(\gamma)$ is independent of the choice of $\tilde{\tau}(s)$. We call $r(\gamma)$ the **rotation index** of γ.

We now prove the theorem of turning tangents:

Theorem 13. *The rotation index of a simple closed curve in E^2 is ± 1.*

Denote by Σ the mapping which sends an ordered pair of points of γ, $\gamma(s_1)$, $\gamma(s_2)$, $0\leqslant s_1\leqslant s_2\leqslant L$ into the end point of the unit vector through O which is parallel to the secant joining $\gamma(s_1)$ to $\gamma(s_2)$. Thus Σ may be considered as mapping the triangle Δ in the (s_1, s_2) plane into the unit circle Γ. The map Σ is continuous, and its restriction to the line $s_1=s_2$ is none other than the tangential mapping T.

Let $p\in\Delta$, and denote by $\tau(p)$ the angle which the axis Ox makes with the line joining O to $\Sigma(p)$, with the restriction that $0\leqslant\tilde{\tau}(p)<2\pi$. As before, we note that $\tau(p)$ is not necessarily a continuous function but nevertheless we can construct a related continuous function $\tilde{\tau}(p)$, $p\in\Delta$ such that $\tilde{\tau}(p)\equiv\tau(p)$ mod 2π. We note that if m is an interior point of Δ, then the argument previously given shows that along each line radiating from m we can construct a function $\tilde{\tau}(p)$ such that $\tilde{\tau}(p)=\tau(p)$ mod 2π and $\tilde{\tau}(p)$ is a function *continous along each radial line from* m. We wish to prove that the function $\tilde{\tau}(p)$ so constructed is *continuous in* Δ.

Let p_o be a point of Δ. Since the segment mp_o is compact and since Σ is continuous, it follows that for any point $q_o\in mp_o$, there is a number η depending only on p_o, such that for any point $q\in\Delta$ whose distance from q_o is less than η, the corresponding points $\Sigma(q)$ and $\Sigma(q_o)$ are not antipodal. Alternatively we have $\tilde{\tau}(q)-\tilde{\tau}(q_o)\not\equiv 0$, mod π.

Now, let ε belong to the open interval $(0, \pi/2)$. Corresponding to this choice of ε, because Σ is continuous, we can choose a neighbourhood U of p_o, such that U is contained in the η-neighbourhood of p_o specified above, and such that, for any $p\in U$, the angle between $O\Sigma(p_o)$ and $O\Sigma(p)$ is less than ε. Alternatively we can write

$$\tilde{\tau}(p)-\tilde{\tau}(p_o)=\varepsilon'+2n(p)\pi\ ,$$

where $|\varepsilon'|<\varepsilon$ and $n(p)$ is an integer.

Let q_o be any point on the segment mp_o, and let q be the point of intersection of the line through q_o parallel to p_op with the line mp. Now the function $\tilde{\tau}(q)-\tilde{\tau}(q_o)$ is continuous in q along mp and is zero when q is at m. Since the distance from q to q_o is less than η, we have $|\tilde{\tau}(q)-\tilde{\tau}(q_o)|<\pi$. In particular when $q_o = p_o$, this gives $|\tilde{\tau}(p) - \tilde{\tau}(p_o)| < \pi$. But this relation with the previous equation gives $n(p) = 0$, showing that $\tilde{\tau}(p)$ is continuous in Δ. In fact, since $\tilde{\tau}(p) \equiv \tau(p) \bmod 2\pi$, it follows that $\tilde{\tau}(p)$ is differentiable in Δ.

If we denote the vertices of Δ by A, B, C where $A = (0, 0)$, $B = (0, L)$, $C = (L, L)$, then the rotation index of γ may be represented by the line integral

$$r = \frac{1}{2\pi} \int_{AC} d\tilde{\tau} \ .$$

This may be decomposed into the sum of two integrals thus:

$$\int_{AC} d\tilde{\tau} = \int_{AB} d\tilde{\tau} + \int_{BC} d\tilde{\tau} \ .$$

We choose a special coordinate system in order to evaluate these two integrals. We can suppose that $\gamma(O)$ is the "lowest" point of γ, and choose this point as the origin O. We choose the "horizontal" tangent vector to γ at O to be the axis Ox. The curve γ then lies in the upper half-plane bounded by Ox. Then the integral $\int_{AB} d\tilde{\tau}$ is equal to the angle rotated by OP as P traverses once along γ. Since OP never points "downwards", this angle must be $\varepsilon\pi$ with $\varepsilon = +1$ or -1. Similarly the integral $\int_{BC} d\tilde{\tau}$ is the angle rotated by PO as P traverses γ and its value is also $\varepsilon\pi$. Thus the sum of the two integrals is $\varepsilon 2\pi$ and the rotation index is ± 1, which completes the proof of the theorem.

An alternative way of writing the rotation index of a closed curve in E^2 is as follows. Using the function $\tilde{\tau}(s)$ defined previously, we see that the components of the unit tangent vector and unit normal may be written

$$e_1 = (\cos \tilde{\tau}(s), \sin \tilde{\tau}(s)) \ ,$$

$$e_2 = (-\sin \tilde{\tau}(s), \cos \tilde{\tau}(s)) \ .$$

Then

$$d\tilde{\tau}(s) = de_1 \cdot e_2 = \kappa ds \ ,$$

from which the rotation index is given by

$$2\pi r(\gamma) = \int_{\gamma} \kappa ds \ .$$

In applications we shall want to consider the rotation index not of a single smooth closed curve but of a sectionally smooth closed curve. Such a curve is made up of a finite number of smooth arcs A_0A_1, A_1A_2, . . . , $A_{m-1}A_m$. The curve is closed if $A_m = A_0$. The idea of rotation index and the theorem of turning tangents can be extended to apply to such curves. Let s_i, $i = 1, 2, \ldots, m$

be the arc length measured from A_0 to A_i. The tangential map is defined at all points except for the vertices $A_0, \ldots, A_m$. At the vertex A_i there are two unit tangent vectors respectively to $A_{i-1}A_i$ and A_iA_{i+1}. The corresponding points on the unit circle Γ are denoted by $T(A_i)^-$ and $T(A_i)^+$. Let ψ_i be the angle from $T(A_i)^-$ to $T(A_i)^+$ such that $0<\psi_i<\pi$— briefly ψ_i is the exterior angle between the tangent to $A_{i-1}A_i$ and A_iA_{i+1}. As in the previous analysis, for each smooth arc $A_{i-1}A_i$ we can construct a continuous function $\tilde{\tau}(s)$ which measures the angle from Ox to the tangent $\gamma'(s)$ at s. The number $r(\gamma)$ defined by

$$2\pi r(\gamma)=\sum_{i=1}^{m}\{\tilde{\tau}(s_i)-\tilde{\tau}(s_{i+1})\}+\sum_{i=1}^{m}\psi_i$$

is an integer called the rotation index of the sectionally smooth closed curve. An argument differing only slightly from the previous one now gives

Theorem 14. *If a sectionally smooth closed curve is simple, then the rotation index is equal to* ± 1.

Exercise

A curve γ in E^2 is said to be **convex** if it lies on one side of every tangent line. Prove that a closed curve with curvature non-negative and rotation index equal to 1 is convex.

2.3 THE GAUSS-BONNET FORMULA

Let us assume, as we can with no loss of generality, that isothermal coordinates are chosen, valid in some neighbourhood D of an oriented surface M. Thus in this neighbourhood, the metric can be written

$$ds^2=e^{2\lambda}(du^2+dv^2) \tag{2.2}$$

where λ is a differentiable function of u and v.

Routine calculation of the Christoffel symbols from this metric gives

$$\Gamma^1_{11}=\Gamma^2_{12}=-\Gamma^1_{22}=\lambda_u\ ,$$
$$\Gamma^1_{12}=\Gamma^2_{22}=-\Gamma^2_{11}=\lambda_v\ ,$$

where we have written $u^1=u$, $u^2=v$, $\lambda_u=\dfrac{\partial\lambda}{\partial u}$, $\lambda_v=\dfrac{\partial\lambda}{\partial v}$. Then a straightforward calculation gives

$$R_{1212}=-e^{2\lambda}(\lambda_{uu}+\lambda_{vv})\ ,$$

so that the Gaussian curvature K is given by

$$K=R_{1212}/g=-e^{-2\lambda}(\lambda_{uu}+\lambda_{vv})\ . \tag{2.3}$$

The element of area of D is given by

$$dA = e^{2\lambda} du dv \,, \tag{2.4}$$

and the area A of D is given by

$$A = \iint_D e^{2\lambda} du dv \,.$$

We now construct a linear differential form ϕ on the 3-dimensional manifold B, the bundle space of the unit tangent bundle over M. Locally a point of B is represented by (u, v, ξ^1, ξ^2) where

$$\xi = \xi^1 \frac{\partial}{\partial u} + \xi^2 \frac{\partial}{\partial v}$$

is a unit tangent vector. Let $\eta = \eta^1 \frac{\partial}{\partial u} + \eta^2 \frac{\partial}{\partial v}$ be the uniquely determined unit tangent vector, orthogonal to ξ such that (ξ, η) is a positive orientation. The differential form ϕ is given by

$$\phi = \langle D\xi, \eta \rangle \,. \tag{2.5}$$

In terms of coordinates, we have

$$\phi = \sum_{i,j=1}^{n} g_{ij} D\xi^i \eta^j \,, \tag{2.6}$$

where

$$D\xi^i = d\xi^i + \sum_{j,k=1}^{n} \Gamma^i_{jk} \xi^j du^k \,. \tag{2.7}$$

The form ϕ given by (2.5) is well-defined in the bundle space B. It is convenient to introduce a coordinate θ defined in the fibre (circle) by

$$\xi^1 = e^{-\lambda} \cos\theta \,, \qquad \xi^2 = e^{-\lambda} \sin\theta \,.$$

Then $\quad \eta^1 = -e^{-\lambda} \sin\theta \,, \quad \eta^2 = e^{-\lambda} \cos\theta \,.$

Substituting these values in (2.6) and (2.7) we get, after simplification,

$$\phi = d\theta - \lambda_v du + \lambda_u dv \,. \tag{2.8}$$

Using (2.3) and (2.4) we find on taking the exterior derivative of (2.8) that

$$d\phi = -K \, dA \,. \tag{2.9}$$

Suppose now that we are given a unit vector field defined over some coordinate neighbourhood N of M, that is, a local cross-section of the bundle B. Then we have an induced 1-form $c^*(\phi)$ on N. Let γ be a smooth curve in N with arc length s and let $\xi(s)$ be a smooth vector field along γ. Then the restriction of $c^*(\phi)$ to γ is of the form σds, and the scalar σ is called the variation of ξ along γ. In particular if ξ is the unit tangent vector field along γ, then σ is the geodesic curvature κ_g of γ. We are now in a position to state and prove the Gauss-Bonnet formula as follows.

Theorem 15. *Let D be a compact oriented domain D in M bounded by a sectionally smooth curve γ. Then*

$$\int_\gamma \kappa_g ds + \int_D K dA + \sum_i (\pi - \alpha_i) = 2\pi\chi \tag{2.10}$$

where κ_g is the geodesic curvature of γ, $\pi - \alpha_i$ are the exterior angles at the vertices of γ, and χ is the Euler characteristic of D.

Proof. We assume first that D belongs to a coordinate neighbourhood (u, v) of M, and that it is bounded by a simple polygon γ of n arcs γ_i, $i = 1, 2, \ldots, n$ with interior angles α_i at the vertices. We assume D positively oriented. To each point of γ_i we associate the unit tangent vector to γ_i. At each vertex of γ there are two tangent vectors making an angle $\pi - \alpha_i$. From theorem 14 it follows that the total variation of θ as the γ_i's are traversed once is $2\pi - \sum_i (\pi - \alpha_i)$. Hence from (2.8)

$$\int_\gamma \kappa_g ds = 2\pi - \sum_i (\pi - \alpha_i) + \int_\gamma (-\lambda_v du + \lambda_u dv)\ .$$

By Stoke's theorem, the last integral is equal to $-\iint K dA$. Moreover, for this special case $\chi(D) = 1$, so that the theorem is proved.

Now assume that D is subdivided into a union of polygons D_λ, $\lambda = 1, 2, \ldots, f$ such that each D_λ lies in one coordinate neighbourhood, and that two different polygons either have no common point, or else they have a vertex or possibly an edge in common. Moreover, we suppose each polygon oriented consistant with that of D so that each interior side has different senses induced from the orientated polygons to which it belongs. Denote by v and e the number of interior vertices and interior edges in the sub-division of D. Then we may apply formula (2.10) to each D_λ and sum. The contributions of the interior edges to the integral involving the geodesic curvature cancel because each edge is traversed in opposite directions. Thus

$$\int_\gamma \kappa_g ds + \iint_D K dA = 2\pi \cdot f - \sum_{i,\lambda} (\pi - \alpha_{i,\lambda}) - \sum_i (\pi - \alpha_i) \tag{2.11}$$

where α_i are the interior angles of D. The sum of all the interior angles at any vertex is 2π. Moreover, each edge is on exactly two of the D_λ. Hence the first term involving summation may be replaced by $-2\pi e + 2\pi v$, and the complete right-hand member becomes $2\pi(v-e+f)-\sum_i(\pi-\alpha_i)$. However we know that $v-e+f=\chi(D)$. Substituting in (2.11) gives the required formula (2.10).

If γ has no vertex, the formula simplifies to

$$\int_\gamma \kappa_g ds + \iint_D KdA = 2\pi\chi \ .$$

In particular, if D is the whole surface M we have

$$\iint_D KdA = 2\pi\chi \ ,$$

and we have proved theorem 11.

As immediate consequences of theorem 11 we have

Corollary 16. *If a closed orientable surface M has zero Gaussian curvature at each point, then M must be homeomorphic to a torus.*

Corollary 17. *If a closed orientable surface M has strictly positive Gaussian curvature at each point, then M must be homeomorphic to a sphere.*

2.4 VECTOR FIELDS AND INDICES

Let D be a domain of a surface M and let ξ be a unit vector field over D with an isolated singularity at some point $p_o \in D$. Let $\gamma(\varepsilon)$ be a geodesic circle, centre p_o and of radius ε. As in the previous section, there is induced from the 1-form ϕ over B a 1-form on this geodesic circle $\gamma(\varepsilon)$ on D. Then the value of the expression

$$\lim_{\varepsilon\to 0}\frac{1}{2\pi}\int_{\gamma(\varepsilon)}\sigma ds$$

is seen from equation (2.8) to be an integer, which is called the **index** of the vector field at p_o.

Examples of vector fields with isolated singularities are shown in Fig. 2.1. The indices are respectively (a) -1; (b) 0; (c) $+1$; (d) $+2$.

We now prove

Theorem 18. *Let M be a closed orientable surface and let ξ be a field of unit tangent vectors over M with a finite number of singularities. Then the sum of the indices is equal to $\chi(M)$.*

To prove the theorem we surround each singularity p_i by a geodesic circle $\gamma_i(\varepsilon)$ of radius ε, and denote by $\Delta_i(\varepsilon)$ the disc bounded by γ_i. We integrate K over the domain $M-\bigcup_i \Delta_i(\varepsilon)$ to get

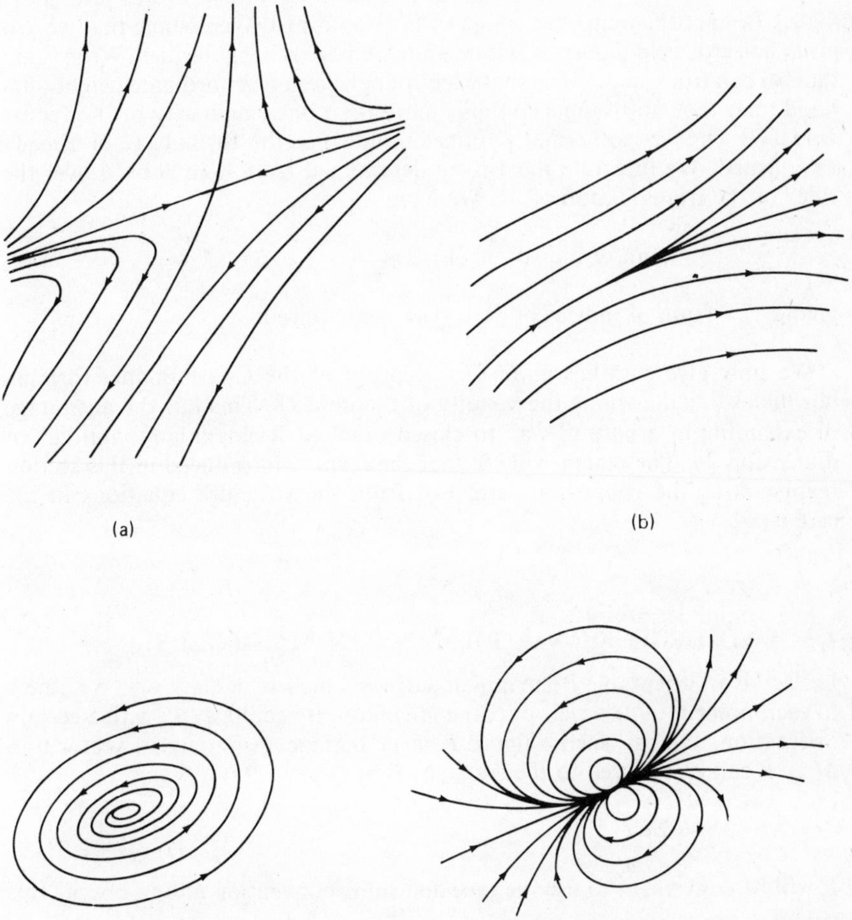

Fig. 2.1

$$\iint_{M-\bigcup_i \Delta_i(\varepsilon)} K dA = \sum_i \int_{\gamma_i(\varepsilon)} \sigma ds \ , \tag{2.11}$$

where $\gamma_i(\varepsilon)$ is oriented as the boundary of $\Delta_i(\varepsilon)$. Now we let $\varepsilon \to 0$ and use the Gauss-Bonnet theorem to get the required result.

We now sketch a direct proof of theorem 18 which avoids using the Gauss-Bonnet theorem, and is due to L. Woodward. We assume that we are given a vector field ξ over M with a finite number of singularities. We assume that we can triangulate M so that each triangle lies in a coordinate neighbourhood, and that no triangle contains more than one singularity of the vector field. We choose isothermal parameters and use the formula (2.8) already established. We integrate the 1-form determined from ϕ in B by ξ over the sides of the triangulation of M. We have

$$2\pi \,(\text{sum of indices of } \xi) = 2\pi(f - e + v) \ ,$$

giving $\quad$ (sum of indices of ξ) $= \chi(M)$ as required.

We now give a rather more formal proof of the Gauss-Bonnet formula in which we shall assume the validity of theorem 18. This has the advantage of extending in a natural way to closed oriented Riemannian manifolds of dimension $2n$. The reader will see that the form ϕ introduced in this section reappears as the connexion form ω_{12} from the structure equations in the case $n = 1$.

2.5 TWO DIMENSIONAL RIEMANNIAN MANIFOLDS

Let M be an orientable Riemannian surface which is of class C^∞. We attach to each point P of M a pair of perpendicular unit vectors e_1, e_2 with a certain orientation. We call such a figure Pe_1e_2 a 2-frame. Any tangent vector v to M at P can be referred to the frame at P by

$$v = \Sigma u_i e_i \ .$$

It will be convenient to use the repeated suffix convention and to rewrite this equation as

$$v = u_i e_i \tag{2.12}$$

where the suffixes run from 1 to 2.

The fundamental equations of Riemannian geometry may be written

$$\left.\begin{aligned} dP &= \omega_i e_i \ , \\ de_i &= \omega_{ij} e_j \ , \\ \omega_{ij} &+ \omega_{ji} = 0 \ , \end{aligned}\right\} \tag{2.13}$$

where ω_i are 1-forms giving a basis dual to the basis (e_i), and where ω_{ij} are the connexion forms.

The equations of structure are

$$\left.\begin{aligned} &d\omega_i = \omega_j \wedge \omega_{ji} \ , \\ &d\omega_{ij} + \omega_{ik} \wedge \omega_{jk} = \Omega_{ij} \ , \\ &\Omega_{ij} + \Omega_{ji} = 0 \ . \end{aligned}\right\} \tag{2.14}$$

The forms Ω_{ij} satisfy a system of equations obtained by taking the exterior derivatives of (2.14). The resulting equations, known as the Bianchi identities are

$$\left.\begin{aligned} &\omega_j \wedge \Omega_{ji} = 0 \ , \\ &d\Omega_{ij} - \omega_{jk} \wedge \Omega_{ik} + \omega_{ik} \wedge \Omega_{jk} = 0 \ . \end{aligned}\right\} \tag{2.15}$$

Suppose now that the frame Pe_1e_2 undergoes a proper orthogonal transformation given by

$$e_i^* = a_{ij}e_j \ , \quad e_i = a_{ji}e_j^* \tag{2.16}$$

where

$$(a_{ij}) = \begin{pmatrix} \cos\theta \ , & -\sin\theta \\ \sin\theta \ , & \cos\theta \end{pmatrix}$$

is an orthogonal 2×2 matrix and θ is a function of the coordinates of P.

It follows from (2.13) and (2.14) that

$$\Omega^*_{ij} = a_{ik}a_{jl}\Omega_{kl} \ . \tag{2.17}$$

We now consider the form Ω defined by

$$4\pi\Omega = -\epsilon_{i_1 i_2}\Omega_{i_1 i_2} \tag{2.18}$$

where $\epsilon_{i_1 i_2}$ is equal to $+1$ if (i_1, i_2) is an even permutation of (1, 2), is equal to -1 if (i_1, i_2) is an odd permutation of (1, 2) and is otherwise zero.

From (2.18) we get

$$\begin{aligned} 4\pi\Omega &= -2\Omega_{12} \\ &= 2R_{1212}\omega_1 \wedge \omega_2 \ , \end{aligned}$$

giving

$$2\pi\Omega = -K\omega_1 \wedge \omega_2 \ , \tag{2.19}$$

where K is the Gaussian curvature of M.

It is clear from (2.17) and (2.19) that Ω remains invariant under a change of frame and is therefore intrinsic. Moreover, the volume element $dA=\omega_1 \wedge \omega_2$ is also invariant under a change of frame. Alternatively we may say that the forms Ω and dA which are defined over the frame bundle are constant on each fibre.

In terms of the notation of this section the Gauss-Bonnet theorem may be written as

$$\int_M -\Omega=\chi(M) \ . \tag{2.20}$$

2.6 THE UNIT TANGENT BUNDLE OF A SURFACE

As local coordinates of B we take local coordinates of M together with the components u_i of (2 12) subject to the condition

$$u_i u_i=1 \ . \tag{2.21}$$

We have

$$dv=\theta_i e_i \tag{2.22}$$

where

$$\theta_i=du_i+u_j\omega_{ji} \ , \tag{2.23}$$

and

$$u_i\theta_i=0 \tag{2.24}$$

because of (2.21) and because ω_{ji} is skew-symmetric.

From (2.23) we get on differentiation

$$\begin{aligned} d\theta_i &= du_j \wedge \omega_{ji}+u_j d\omega_{ji} \\ &= \theta_j \wedge \omega_{ji}-u_k\omega_{kj} \wedge \omega_{ji}+u_k(\Omega_{ki}-\omega_{kl} \wedge \omega_{il}) \ , \end{aligned}$$

that is

$$d\theta_i=\theta_j \wedge \omega_{ji}+u_j\Omega_{ji} \ . \tag{2.25}$$

A change of frame (2.16) leads to the following change of components u_i, θ_i given by

$$u_i^*=a_{ij}u_j \ , \quad \theta_i^*=a_{ij}\theta_j \ . \tag{2.26}$$

We now construct differential forms Φ_0 of degree 1 and Ψ_0 of degree 2:

$$\Phi_0=\epsilon_{i_1i_2}u_{i_1}\theta_{i_2} \ , \tag{2.27}$$

$$\Psi_0=\epsilon_{i_1i_2}\Omega_{i_1i_2} \ . \tag{2.28}$$

Again we see that Φ_0 and Ψ_0 are invariant over a change of frame, and hence they are intrinsic forms defined globally over the unit tangent bundle B. We have

$$\begin{aligned} d\Phi_0 &= \epsilon_{i_1i_2} du_{i_1} \wedge \theta_{i_2} + \epsilon_{i_1i_2} u_{i_1} d\theta_{i_2} \\ &= \epsilon_{i_1i_2}(\theta_{i_1} \wedge \theta_{i_2} - u_j\omega_{ji_1} \wedge \theta_{i_2}) + \epsilon_{i_1i_2} u_{i_1}(\theta_j \wedge \omega_{ji_2} + u_j\Omega_{ji_2}) \ . \end{aligned}$$

This expression consists of terms involving ω_{ij} and those which do not. The terms not involving ω_{ij} make up the intrinsic expression

$$\begin{aligned} \epsilon_{i_1i_2}\theta_{i_1} \wedge \theta_{i_2} + \epsilon_{i_1i_2} u_{i_1} u_j \Omega_{ji_2} &= 2\theta_1 \wedge \theta_2 + u_1 u_j \Omega_{j2} - u_2 u_j \Omega_{j1} \\ &= 2\theta_1 \wedge \theta_2 + (u_1{}^2 + u_2{}^2)\Omega_{12} \\ &= 2\theta_1 \wedge \theta_2 + \Omega_{12} \ . \end{aligned} \tag{2.29}$$

From (2.24) it follows that θ_1, θ_2 are linearly dependent, so this expression reduces to Ω_{12} .

Now we have already seen in chapter 1 that we can always choose local normal coordinates in a neighbourhood of the point P so that the Christoffel symbols vanish at P; equivalently, at P, we have $\omega_{ij}=0$. Since the relation is between intrinsic properties we have proved the important result

$$d\Phi_0 = \tfrac{1}{2}\Psi_0 \ . \tag{2.30}$$

Now we define on M a continuous field of unit tangent vectors with a point O of M as the only singularity. We now make use of theorem 12 (which we have proved independently of the Gauss-Bonnet theorem) which asserts that the index of the vector field at O is the Euler characteristic $\chi(M)$. This vector field defines in B a submanifold V which has as boundary χZ where Z is the 1-dimensional cycle formed by all the unit tangent vectors through O. The integral of Ω over M is clearly equal to the integral of Ω over V. Using Stoke's theorem we get

$$\int_M \Omega = \int_V \Omega = -\frac{1}{4\pi} \int_V \Psi_0 = -\frac{1}{2\pi} \int_V d\Phi_0 = -\frac{\chi}{2\pi} \int_Z \Phi_0 \ . \tag{2.31}$$

But $\Phi_0 = u_1\theta_2 - u_2\theta_1$ is the volume element of the 1-dimensional unit sphere. Therefore,

$$\int_Z \Phi_0 = 2\pi \ .$$

Thus we have proved

$$\int_M -\Omega = \chi(M) \ , \tag{2.32}$$

that is

$$\frac{1}{2\pi}\int_M KdA = \chi(M) \ . \tag{2.33}$$

The reader may be puzzled why we have given another proof of the Gauss-Bonnet Theorem which seems more complicated than the previous one. The reason is that a proof precisely similar to the latter can be extended from 2-dimensional to $2p$-dimensional Riemannian manifolds, and we have really prepared the way for this extension which is treated in the next section.

2.7 THE 2p-DIMENSIONAL CASE

Let M be an orientable Riemannian manifold of even dimension, $n=2p$. We attach to each point of M an orthonormal frame of tangent vectors $e_1, e_2, \ldots, e_n$ with a certain orientation. The argument employed in section 2.5 goes over in this more general case as far as equation (2.17), where suffixes i, j now take values $1, 2, \ldots, n$.

Instead of (2.18) we consider the form Ω defined by

$$\Omega = (-1)^p \frac{1}{2^{2p}\pi^p p!} \epsilon_{i_1 \ldots i_{2p}} \Omega_{i_1 i_2}\Omega_{i_3 i_4} \ldots \Omega_{i_{2p-1} i_{2p}} \ . \tag{2.34}$$

From (2.17) it follows that Ω remains invariant under a change of frame (2.16) and is therefore intrinsic. We shall write the Gauss-Bonnet formula in the form

$$\int_{M^n} -\Omega = \chi \ . \tag{2.35}$$

The space of unit tangent vectors over M forms a manifold of dimension $2n-1$. Equations (2.21), (2.22), (2.23), (2.24), (2.25), (2.26) are still valid in this more general case. However the technical details become much more complicated at this stage.

We now construct the following two sets of differential forms:

$$\Phi_k = \epsilon_{i_1 \ldots i_{2p}} u_{i_1} \theta_{i_2} \ldots \theta_{i_{2p-2k}} \Omega_{i_{2p-2k+1} i_{2p-2k+2}} \ldots \Omega_{i_{2p-1} i_{2p}} \ , \quad k = 0, 1, \ldots, p-1 \ , \tag{2.36}$$

$$\Psi_k = \epsilon_{i_1 \ldots i_{2p}} \Omega_{i_1 i_2} \theta_{i_3} \ldots \theta_{i_{2p-2k}} \Omega_{i_{2p-2k+1} i_{2p-2k+2}} \ldots \Omega_{i_{2p-1} i_{2p}} \quad k = 0, 1, \ldots, p-1 \ . \tag{2.37}$$

The forms Φ_k are of degree $2p-1$ and Ψ_k of degree $2p$, and we remark that Ψ_{p-1} differs from Ω only by a numerical factor. Using (2.17) and (2.26), we see that Φ_k and Ψ_k are intrinsic and are therefore defined over the entire Riemannian manifold M.

Theorem 19. *The following recurrent relation holds*

$$d\Phi_k = \Psi_{k-1} + \frac{2p-2k-1}{2(k+1)} \Psi_k\ , \quad k=0,1,\ldots,p-1\ , \tag{2.38}$$

where we define $\Psi_{-1}=0$.

Using the property of skew-symmetry of the symbol $\epsilon_{i_1 \ldots i_{2p}}$ in its indices, we can write

$$\begin{aligned}d\Phi_k &= \epsilon_{(i)} du_{i_1}\theta_{i_2} \ldots \theta_{i_{2p-2k}}\Omega_{i_{2p-2k+1}i_{2p-2k+2}} \ldots \Omega_{i_{2p-1}i_{2p}}\\ &+(2p-2k-1)\epsilon_{(i)}u_{i_1}d\theta_{i_2}\theta_{i_3} \ldots \theta_{i_{2p-2k}}\Omega_{i_{2p-2k+1}i_{2p-2k+2}} \ldots \Omega_{i_{2p-1}i_{2p}}\\ &-k\epsilon_{(i)}u_{i_1}\theta_{i_2} \ldots \theta_{i_{2p-2k}}d\Omega_{i_{2p-2k+1}i_{2p-2k+2}}\Omega_{i_{2p-2k+3}i_{2p-2k+4}} \ldots \Omega_{i_{2p-1}i_{2p}}\ ,\end{aligned}$$

where $\epsilon_{(i)}$ is an abbreviation of $\epsilon_{i_1 \ldots i_{2p}}$. For the derivatives du_i, $d\theta_i$, $d\Omega_{ij}$ we can substitute their expressions from (2.23), (2.25) and (2.15). The resulting expression for $d\Phi_k$ will then consist of terms of two kinds, those involving ω_{ij} and those not. We collect the terms not involving ω_{ij}, which are

$$\Psi_{k-1} + (2p-2k-1)\epsilon_{(i)}u_{i_1}u_j\Omega_{ji_2}\theta_{i_3} \ldots \theta_{i_{2p-2k}}\Omega_{i_{2p-2k+1}i_{2p-2k+2}} \ldots \Omega_{i_{2p-1}i_{2p}}\ . \tag{2.39}$$

This expression is obviously intrinsic. Its difference with $d\Phi_k$ is an expression which contains a factor ω_{ij} in each of its terms.

As before this difference is shown to be zero by choosing normal co-ordinates based at an arbitrary point P so that, at P,

$$\omega_{ij}=0\ .$$

To transform the expression (2.39) we shall introduce the abbreviations

$$\begin{aligned}P_k &= \epsilon_{(i)}u_{i_1}^2\Omega_{i_1i_2}\theta_{i_3} \ldots \theta_{i_{2p-2k}}\Omega_{i_{2p-2k+1}i_{2p-2k+2}} \ldots \Omega_{i_{2p-1}i_{2p}}\ ,\\ \Sigma_k &= \epsilon_{(i)}u_{i_1}u_{i_3}\Omega_{i_3i_2}\theta_{i_3} \ldots \theta_{i_{2p-2k}}\Omega_{i_{2p-2k+1}i_{2p-2k+2}} \ldots \Omega_{i_{2p-1}i_{2p}}\ ,\\ T_k &= \epsilon_{(i)}u_{i_3}^2\Omega_{i_1i_2}\theta_{i_3} \ldots \theta_{i_{2p-2k}}\Omega_{i_{2p-2k+1}i_{2p-2k+2}} \ldots \Omega_{i_{2p-1}i_{2p}}\ ,\end{aligned} \tag{2.40}$$

which are forms of degree $2p$. Owing to the relations (2.21) and (2.23) there are some simple relations between these forms and Ψ_k. In fact, we can write

$$\begin{aligned}P_k &= \epsilon_{(i)}(1-u_{i_2}^2-u_{i_3}^2- \ldots -u_{i_{2p}}^2)\Omega_{i_1i_2}\theta_{i_3} \ldots \theta_{i_{2p-2k}}\Omega_{i_{2p-2k+1}i_{2p-2k+2}} \ldots\\ &\quad \ldots \Omega_{i_{2p-1}i_{2p}}\\ &= \Psi_k - P_k - 2(p-k-1)T_k - 2kP_k\ ,\end{aligned}$$

which gives

$$\Psi_k = 2(k+1)P_k + 2(p-k-1)T_k\ . \tag{2.41}$$

Again, we have

$$\begin{aligned}\Sigma_k = {} & \epsilon_{(i)} u_{i_1} \Omega_{i_3 i_2}(-u_{i_1}\theta_{i_1} - u_{i_2}\theta_{i_2} - u_{i_4}\theta_{i_4} - \ldots - u_{i_{2p}}\theta_{i_{2p}})\theta_{i_4} \ldots \\ & \ldots \theta_{i_{2p-2k}} \Omega_{i_{2p-2k+1} i_{2p-2k+2}} \ldots \Omega_{i_{2p-1} i_{2p}} \\ = {} & T_k - (2k+1)\Sigma_k \ ,\end{aligned}$$

and hence

$$T_k = 2(k+1)\Sigma_k \ . \tag{2.42}$$

The expression (2.39) for $d\Phi_k$ therefore becomes

$$d\Phi_k = \Psi_{k-1} + (2p-2k-1)\{P_k + 2(p-k-1)\Sigma_k\} \ , \quad k = 0, 1, \ldots, p-1 \ .$$

Using (2.41) and (2.42), we get the desired formula (2.38).

From (2.38) we can solve Ψ_k in terms of $d\Phi_0, d\Phi_1, \ldots, d\Phi_k$. The result is found to be

$$\Psi_k = \sum_{m=0}^{k} (-1)^m \frac{2^{m+1}(k+1)k \ldots (k-m+1)}{(2p-2k-1)(2p-2k+1) \ldots (2p-2k+2m-1)} d\Phi_{k-m} \ , \qquad k = 0, 1, \ldots, p-1 \ . \tag{2.43}$$

In particular, it follows that Ω is the exterior derivative of a form Π:

$$\Omega = (-1)^p \frac{1}{2^{2p}\pi^p p!} \Psi_{p-1} = d\Pi \ , \tag{2.44}$$

where

$$\Pi = \frac{1}{\pi^p} \sum_{m=0}^{p-1} (-1)^{m+1} \frac{1}{1.3 \ldots (2p-2m-1) m! 2^{p+m}} \Phi_m \ . \tag{2.45}$$

As in the 2-dimensional case we consider a continuous field of unit tangent vectors over M^n with the point O as the only singular point. Again we appeal to Hopf's theorem that the index of the field at O is equal to the Euler-Poincaré characteristic of M. But now the vector field defines in M^{2n-1} a submanifold V^n, with boundary χZ where Z is the $(n-1)$-dimensional cycle formed by all unit vectors through O. The integral of Ω over M^n is equal to the integral over V^n. Using Stoke's theorem we get

$$\int_{M^n} \Omega = \int_{V^n} \Omega = \chi \int_Z \Pi = -\chi \frac{1}{1.3 \ldots (2p-1) 2^p \pi^p} \int_Z \Phi_0 \ . \tag{2.46}$$

Now from the definition of Φ_0 we have

$$\Phi_0 = (2p-1)! \sum_{i=1}^{n} (-1)^i \theta_1 \ldots \theta_{i-1} u_i \theta_{i+1} \ldots \theta_{2p} \ . \tag{2.47}$$

Let dV denote the volume element of the $(2p-1)$-dimensional unit sphere. We see from equation (2.21) that the projection of this element on the various hyperplanes is given, for example, by

$$u^i dV = (-1)^i \theta_1 \ldots \theta_{i-1} \hat{\theta}_i \theta_{i+1} \ldots \theta_{2p} \tag{2.48}$$

where the symbol ^ denotes that this element is deleted.

Multiply (2.48) by u^i and sum with respect to i. Since $u^i u^i = 1$ we get

$$dV = \sum_{i=1}^{n} (-1)^i \theta_1 \ldots \theta_{i-1} u_i \theta_{i+1} \ldots \theta_{2p}$$

from which we deduce that Φ_0 is none other than $(2p-1)!$ times the volume element of the $(2p-1)$-dimensional sphere. Therefore

$$\int \Phi_0 = (2p-1)! \frac{2\pi^p}{(p-1)!} .$$

Using this in (2.46) gives the required formula (2.35), and the generalized Gauss-Bonnet theorem has been proved.

2.8 SPECIAL CASES

One of the most interesting unresolved conjectures in Riemannian geometry is the following:

> *Let M be a compact oriented Riemannian manifold of even dimension $n = 2p$. If all the sectional curvatures of M are non-negative then the Euler characteristic $\chi(M) \geqslant 0$. If all the sectional curvatures of M are non-positive, then $(-1)^p \chi(M) \geqslant 0$.*

The conjecture is easily proved if dim $M = 2$. The case dim $M = 4$ has also been proved by J. Milnor. In fact, Milnor showed that the weaker hypothesis that the Ricci curvature being positive is sufficient to prove that $\chi(M) \geqslant 0$. [R. L. Bishop and S. I. Goldberg, *Some implications of the generalized Gauss-Bonnet theorem*, Trans. Amer. Math. Soc. 112 (1964), 508–535, MR 29 # 574; S. S. Chern, *On curvature and characteristic classes of a Riemannian manifold*, Abh. Math. Sem. Univ. Hamburg 20 (1956), 117–126, MR 17, # 783.]

Another result for 4-dimensional manifolds is due to A. Avez [*Applications de la formule de Gauss-Bonnet-Chern aux variétés à quatre dimensions*, C. R. Acad. Sci. Paris, 256 (1963), 5488–5490].

He showed that the Gauss-Bonnet formula can be written in the form

$$\int_M ||R||^2 dv = \int_M ||\rho - \frac{\tau}{4} g||^2 dv + 32\pi^2 \chi(M) .$$

From the formula we have

Theorem 20. *A four-dimensional torus with an Einstein metric is necessarily flat.*

This follows because $\chi(M)=0$ and $\rho=\frac{\tau}{4}g$ yields $||R||^2=0$ from which the result follows.

In the 6-dimensional case, the Gauss-Bonnet formula may be written as

$$\chi(M)=\frac{1}{8\pi^2}\int_M \Omega_{123456} ,$$

where Ω_{123456} is defined as follows:
Let Ω_{ij} be the curvature form of M. Let

$$\Omega_{ijkl}=\Omega_{ij}\wedge\Omega_{kl}-\Omega_{ik}\wedge\Omega_{jl}+\Omega_{il}\wedge\Omega_{jk} .$$

Then

$$\Omega_{123456}=\Omega_{12}\wedge\Omega_{3456}-\Omega_{13}\wedge\Omega_{2456}+\Omega_{14}\wedge\Omega_{2356}-\Omega_{15}\wedge\Omega_{2346} +\Omega_{16}\wedge\Omega_{2345} .$$

However, it seems difficult to make further progress in the 6-dimensional case unless one makes further restrictions upon the metric.

2.9 CURVATURE INVARIANTS AND THE GAUSS-BONNET THEOREM

A scalar-valued curvature invariant is a polynomial in the components of the curvature tensor and its covariant derivatives which is independent of the choice of basis of the tangent space of M at each point m. In fact, it follows from Weyl's theory of invariants that the invariant polynomials are contractions of the components of the curvature tensor and its covariant derivatives. An invariant is said to have order $2k$ if it involves a total of $2k$ derivatives of the metric tensor.

It is known that if $I(k, n)$ denotes the vector space of invariants of order $2k$, then $I(1, n)$ has dimension 1 for $n\geqslant 2$, and dim $I(2, n)=4$ for $n\geqslant 4$. In fact, writing

$$\tau=\Sigma R_{ijij} , \quad ||\rho||^2=\Sigma\rho_{ij}^2 , \quad ||R||^2=\Sigma R_{ijkl}^2 , \quad \Delta\tau=\Sigma\nabla_{ii}^2\tau ,$$

we find that $\{\tau\}$ is a basis for $I(1, n)$ and that $\{\tau^2, ||\rho||^2, ||R||^2, \Delta\tau\}$ is a basis for $I(2, n)$.

It is known that dim $I(3, n)=17$ for $n\geqslant 6$. However, the 6-dimensional Gauss-Bonnet integrand must vanish for manifolds of dimension $\leqslant 5$. In terms of the order 6 invariants introduced in chapter 1 this leads to the following identity:

$$\tau^3 + 3\tau||R||^2 - 12\tau||\rho||^2 + 16\check{\rho} + 4\check{R} - 8\check{\bar{R}} \\ + 24 < \rho \otimes \rho, \bar{R} > - 24 < \rho\ , \dot{R} > = 0\ . \tag{2.49}$$

This is an unexpected pay-off from the Gauss-Bonnet theorem.

2.10 THE SIGN OF THE GAUSS-BONNET INTEGRAL

If the Gauss-Bonnet integral is strictly positive at all points, then the Euler-Poincaré characteristic is strictly positive. A natural problem is to impose conditions on the Riemann curvature tensor which will ensure that this is satisfied. A rather surprising result was obtained by R. Geroch (1974) who showed that there exist examples of Riemannian manifolds with strictly positive sectional curvature for which the Gauss-Bonnet integrand assumes negative values at certain points. Analogous examples were subsequently given by P. Klembeck (1976).

Associated with the Riemann tensor R is the so-called curvature operator $\hat{R}$ which maps the space of skew-symmetric covariant tensors at $p \in M$ to itself. Equally well, we may regard $\hat{R}$ as mapping the space of bi-vectors at p to itself. Let (e_i) be an orthonormal basis for the tangent vectors at p, and let a, b be tangent vectors at p. Let ω be a bi-vector at p. Then we define

$$\hat{R}(\omega)(a, b) = \sum_{i,j=1}^{n} R(e_i, e_j, a, b)\omega(e_i, e_j)\ . \tag{2.50}$$

R. S. Kulkarni (1972) has shown that by requiring $\hat{R}$ to be a positive definite linear operator on the space of bi-vectors, we can guarantee that the Gauss-Bonnet integrand is positive.

An extremely interesting survey and development of these ideas is to be found in the paper by J. P. Bourguignon and H. Karcher (1978).

2.11 EXERCISES

(1) Let ABC be a geodesic triangle on a surface S which contains a point P and let K be the Gaussian curvature of the surface at P. Let Δ be the area of the triangle ABC. Prove that

$$K = \lim\,(A + B + C - \pi)/\Delta$$

where the limit is taken as all vertices tend to P.

(2) An anchor ring is obtained by rotating a circle of radius a about a line in its plane and at distance b $(>a)$ from its centre. Its equations are given by

$$x = (b + a\cos u)\cos v\ ,$$
$$y = (b + a\cos u)\sin v\ ,$$
$$z = a\sin u\ ,$$

where $0<u<2\pi$, $0<v<2\pi$.

Find the Gaussian curvature at the point (u, v), and verify that the total curvature of the whole surface is zero.

(3) By considering the region of the anchor ring of exercise 2 bounded by two meridians and two parallels $u=0$, $u=\pi$, prove that the total curvature of the whole surface is zero.

(4) The vertex of a right circular cone of semi-vertical angle α is smoothed so that there is no longer a singularity on the surface. Prove that the total curvature of the surface is increased by $2\pi(1-\sin\alpha)$.

(5) Prove that the total curvature of a paraboloid of revolution is 2π. Find the total curvature of one sheet of a two-sheeted hyperboloid of revolution.

(6) Prove that on a general compact 4-dimensional orientable Riemannian manifold M we have

$$\int_M (||R||^2-4||\rho||^2+\tau^2)dV=32\pi^2\chi$$

where χ is the Euler-characteristic of M.
[For a detailed solution see Avez (1963).]

Chapter 3

Total Absolute Curvature

3.1 INTRODUCTION

In the previous chapter we were concerned with the intrinsic curvature measures of a Riemannian manifold. In this chapter we shall consider extrinsic curvatures, that is, invariants which depend on the way in which a manifold is immersed in another manifold. A good text on the geometry of submanifolds is to be found in the book by Bang-yen-Chen, *Geometry of Submanifolds*, Marcel Dekker, New York, 1973, and our treatment follows closely certain sections of this book.

Let M, N be differentiable manifolds of dimensions m and n respectively with $m<n$, and let $f\colon M\to N$ be an immersion of M into N. Let N carry a Riemannian metric $\tilde{g}$. Then f induces a Riemannian metric g on M. Let X, Y be two tangent vectors at P of M. Then g is determined by the formula

$$g_p(X, Y)=\tilde{g}_{f(P)}(f_*X, f_*Y) \ ,$$

this making sense since we may regard f as locally injective. In fact, the image of the tangent space to M at P is an m-dimensional subspace of the n-dimensional tangent space to N at $f(P)$.

A tangent vector ξ to N at $f(P)$ is called **normal** if

$$\tilde{g}_{f(P)}(f_*X, \xi)=0$$

for any tangent vector X of M at P. A set of such vectors, one at each point $f(P)$, is called a **normal vector field**. Let $T^{\perp}(M)$ denote the vector bundle of all normal vectors. Then the tangent bundle of N restricted to the image of M is isomorphic to the direct sum of the tangent bundle $T(M)$ of M and the normal bundle $T^{\perp}(M)$.

Let X be a vector field on M. Assuming that f is locally injective, the image $f_*(X)$ will be a vector field on its image on N. However, we wish to make use of vector fields which are defined over a neighbourhood of N and not restricted to the image of M. Clearly there are many ways of extending the vector field $f_*(X)$ to such a neighbourhood, and we denote such an extension by $\tilde{X}$. It turns out, however, that important invariants constructed from such extensions when themselves restricted to the image of M, are independent of the particular extensions chosen.

In order to illustrate such an extension, let us represent the map f locally in terms of local coordinates u^A in N, $(A=1, 2, \ldots, n)$ and local coordinates x^i in M $(i=1, 2, \ldots, m)$. We write

$$u^A=f^A(x^i) \ ,$$

or, by an abuse of notation which will clearly cause no confusion,

$$u^A=u^A(x^i) \ .$$

The vector field $X=X^h\dfrac{\partial}{\partial x^h}$ will be mapped by f_* into the vector

$$\left(\frac{\partial u^A}{\partial x^h}X^h\right)\frac{\partial}{\partial u^A}$$

on N. Here we are using the summation convention of summing over repeated indices. Let $\tilde{X}(u)$ denote some extension of $f_*(X)$ to the coordinate neighbourhood of N. Then the restriction of $\tilde{X}(u)$ to points $u(x)$ must clearly be $f_*(X)$, that is,

$$\tilde{X}^A(u(x))=\frac{\partial u^A}{\partial x^h}X^h(u) \ .$$

Let $\tilde{Y}$ be some extension of $f_*(Y)$ where Y is a vector field on M, so that

$$\tilde{Y}^A(u(x))=\frac{\partial u^A}{\partial x^h}Y^h(u) \ .$$

Lemma 21. *Let $\tilde{X}$, $\tilde{Y}$ be extensions of f_*X, f_*Y respectively. Then*

$$[\tilde{X}, \tilde{Y}]_{|f(M)}=[f_*X, f_*Y]=f_*[X, Y] \ .$$

Hence the restriction of $[\tilde{X}, \tilde{Y}]$ to $f(M)$ is independent of the particular extensions $\tilde{X}$, $\tilde{Y}$ which are chosen.

To prove the lemma, we note that

$$[\tilde{X}, \tilde{Y}]^A_{|u=u(x)}=\left(\tilde{X}^B\frac{\partial \tilde{Y}^A}{\partial u^B}-\tilde{Y}^B\frac{\partial \tilde{X}^A}{\partial u^B}\right)_{|u=u(x)}$$

$$=\left(\frac{\partial u^B}{\partial x^h}X^h\frac{\partial \tilde{Y}^A}{\partial u^B}-\frac{\partial u^B}{\partial x^h}Y^h\frac{\partial \tilde{X}^A}{\partial u^B}\right)_{|u=u(x)}$$

$$=X^h\frac{\partial}{\partial x^h}\left(\frac{\partial u^A}{\partial x^i}Y^i\right)-Y^h\frac{\partial}{\partial x^h}\left(\frac{\partial u^A}{\partial x^i}X^i\right)$$

$$=\frac{\partial u^A}{\partial x^i}\left(X^h\frac{\partial Y^i}{\partial x^h}-Y^h\frac{\partial X^i}{\partial x^h}\right)$$
$$=(f_*[X,Y])^A .$$

It follows that the restriction of $[\tilde{X}, \tilde{Y}]$ to $f(M)$ is independent of the particular extensions $\tilde{X}$, $\tilde{Y}$ which have been chosen.

We now prove

Lemma 22. *Let X, Y be two vector fields on M and let $\tilde{X}$, $\tilde{Y}$ be extensions of $f_*(X), f_*(Y)$ respectively. Then the restriction of $\tilde{\nabla}_{\tilde{X}}\tilde{Y}$ to $f(M)$ does not depend on the particular extensions chosen.*

To prove the lemma, we have

$$(\tilde{\nabla}_{\tilde{X}}\tilde{Y})^C=\tilde{X}^B(\partial_B\tilde{Y}^C+\left\{{C \atop AB}\right\}\tilde{Y}^A) .$$

Thus

$$(\tilde{\nabla}_{\tilde{X}}\tilde{Y})_{|u=u(x)}=\frac{\partial u^B}{\partial x^i}X^i\left(\partial_B\tilde{Y}^C+\left\{{C \atop AB}\right\}\frac{\partial u^A}{\partial x^j}Y^j\right)_{|u=u(x)}$$
$$=X^i\left(\partial_i\left(\frac{\partial u^C}{\partial x^j}Y^j\right)+\left\{{C \atop AB}\right\}\frac{\partial u^B}{\partial x^i}\frac{\partial u^A}{\partial x^j}Y^j\right) .$$

Therefore $\tilde{\nabla}_{\tilde{X}}\tilde{Y}_{|f(M)}$ does not depend on the choice of extensions, and we can denote this expression by $\tilde{\nabla}_X Y$. Here we have effectively identified X with $f_*(X)$ and Y with $f_*(Y)$. We do this to avoid using an unwieldy notation, and do not believe that any confusion will arise.

3.1.1 The second fundamental form

We now prove

Lemma 23. *The vector field $\tilde{\nabla}_X Y$ may be decomposed into a part $T_X Y$ tangential to $f(M)$ and a part $h(X, Y)$ normal to $f(M)$. Moreover $T_X Y$ is none other than $\nabla_X Y$ where ∇ is the Riemannian connexion of M with respect to the induced metric g. Furthermore the function h is bilinear and symmetric in X and Y.*

We write

$$\tilde{\nabla}_X Y=T_X Y+h(X, Y) . \qquad (3.1)$$

Then, for functions α, β on M we have

$$\tilde{\nabla}_{\alpha X}(\beta Y)=\alpha\{(X\beta)Y+\beta(\tilde{\nabla}_X Y)\}$$
$$=\alpha(X\beta)Y+\alpha\beta(T_X Y+h(X, Y)) . \qquad (3.2)$$

But

$$\tilde{\nabla}_{\alpha X}(\beta Y) = T_{\alpha X}(\beta Y) + h(\alpha X, \beta Y) . \tag{3.3}$$

Equate the normal components in (3.2) and (3.3) to get

$$h(\alpha X, \beta Y) = \alpha\beta h(X, Y) .$$

Since the additivity property is trivial, it follows that h is bilinear.

Equate the tangential components in (3.2) and (3.3) to get

$$T_{\alpha X}(\beta Y) = \alpha(X\beta)Y + \alpha\beta T_X Y .$$

Thus T determines an affine connexion which we denote by ∇, so (3.1) can be rewritten as

$$\tilde{\nabla}_X Y = \nabla_X Y + h(X, Y) . \tag{3.4}$$

Since $\tilde{\nabla}$ has zero torsion it follows that

$$\tilde{\nabla}_X Y - \tilde{\nabla}_Y X - [X, Y] = 0 .$$

Hence

$$\nabla_X Y + h(X, Y) - \nabla_Y X - h(Y, X) - [X, Y] = 0 .$$

Equating to zero the tangential part of the left-hand member gives

$$\nabla_X Y - \nabla_Y X - [X, Y] = 0 , \tag{3.5}$$

thus proving that ∇ has zero torsion.

Equating to zero the normal part of the left-hand member gives

$$h(X, Y) = h(Y, X)$$

thus proving that h is symmetric

Moreover,

$$\begin{aligned}\nabla_X(g(Y, Z)) &= \tilde{\nabla}_X(\tilde{g}(Y, Z))\\ &= \tilde{g}(\tilde{\nabla}_X Y, Z) + \tilde{g}(Y, \tilde{\nabla}_X Z)\\ &= \tilde{g}(\nabla_X Y + h(X, Y), Z) + \tilde{g}(Y, \nabla_X Z + h(X, Z))\\ &= \tilde{g}(\nabla_X Y, Z) + \tilde{g}(Y, \nabla_X Z)\\ &= g(\nabla_X Y, Z) + g(Y, \nabla_X Z) ,\end{aligned}$$

thus proving that ∇ is a metric connexion. This, together with (3.5), shows that ∇ is just the Riemannian connexion of g.

The bilinear form h is called the **second fundamental form** of the submanifold M. For each point p of M, the value of $h(X, Y)$ depends only on X_p and Y_p.

3.1.2 The normal connexion

We now show how another connexion called the **normal connexion** is determined by our configuration. Let ξ be a normal vector field to M. Then the vector field $\tilde{\nabla}_X\xi$ may be decomposed into its tangential part $-A_\xi X$ and its normal part $\overset{\perp}{\nabla}_X\xi$ as follows:

$$\tilde{\nabla}_X\xi = -(A_\xi X) + \overset{\perp}{\nabla}_X\xi \ . \tag{3.6}$$

We now prove

Lemma 24. *The expression $A_\xi X$ is bilinear in ξ and X, and as a consequence its value at p depends only on ξ_p and X_p. Moreover*

$$g(A_\xi X, Y) = \tilde{g}(h(X, Y), \xi) \tag{3.7}$$

for any vector fields X, Y on M.

To prove the lemma, let α, β be any two functions on M. Then

$$\begin{aligned}\tilde{\nabla}_{\alpha X}(\beta\xi) &= \alpha\{(X\beta)\xi + \beta\tilde{\nabla}_X\xi\} \\ &= \alpha(X\beta)\xi - \alpha\beta A_\xi X + \alpha\beta\overset{\perp}{\nabla}_X\xi \ .\end{aligned}$$

Also

$$\tilde{\nabla}_{\alpha X}(\beta\xi) = -A_{\beta\xi}(\alpha X) + \overset{\perp}{\nabla}_{\alpha X}(\beta\xi) \ .$$

Equating tangential parts gives

$$A_{\beta\xi}(\alpha X) = \alpha\beta A_\xi X \ . \tag{3.8}$$

Since the additivity property is trivial, it follows from (3.8) that $A_\xi X$ is bilinear.

To prove (3.7) we remark that, for an arbitrary vector field Y on M,

$$\begin{aligned}0 &= \tilde{g}(\tilde{\nabla}_X Y, \xi) + \tilde{g}(Y, \tilde{\nabla}_X\xi) \\ &= \tilde{g}(\nabla_X Y + h(X, Y), \xi) + \tilde{g}(Y, -A_\xi X + \overset{\perp}{\nabla}_X\xi) \\ &= \tilde{g}(h(X, Y), \xi) - \tilde{g}(Y, A_\xi X) \ ,\end{aligned}$$

from which the result follows.

If we now equate normal parts we find, analogous to (3.8),

$$\overset{\perp}{\nabla}_{\alpha X}(\beta\xi) = \alpha(X\beta)\xi + \alpha\beta\overset{\perp}{\nabla}_X\xi \ . \tag{3.9}$$

This equation justifies our notation $\overset{\perp}{\nabla}$ since this is essentially the defining property of a connexion. Indeed, $\overset{\perp}{\nabla}$ is called the **normal connexion.**

Lemma 25. *$\overset{\perp}{\nabla}$ is a metric connexion in the normal bundle M in N.*

Let ξ, η be two normal vector fields. Then

$$\tilde{\nabla}_X\xi = -A_\xi X + \overset{\perp}{\nabla}_X\xi \ , \quad \tilde{\nabla}_X\eta = -A_\eta X + \overset{\perp}{\nabla}_X\eta \ .$$

Hence

$$\begin{aligned} \tilde{g}(\overset{\perp}{\nabla}_X\xi, \eta) + \tilde{g}(\xi, \overset{\perp}{\nabla}_X\eta) &= \tilde{g}(\tilde{\nabla}_X\xi, \eta) + \tilde{g}(\xi, \tilde{\nabla}_X\eta) \\ &= \tilde{\nabla}_X\tilde{g}(\xi, \eta) \\ &= \overset{\perp}{\nabla}_X\tilde{g}(\xi, \eta) \ , \end{aligned}$$

and the lemma is proved.

3.1.3 The canonical metric on the normal and unit normal bundles

We note that under parallel transport with respect to $\overset{\perp}{\nabla}$, a unit normal vector is transported into a unit normal vector.

We now prove a result which will have important applications. The essential point is that the normal bundle has induced on it canonically a Riemannian metric determined by the immersion $f: M \to N$.

The normal bundle is an n-dimensional differentiable manifold—the base space has dimension m and the fibre dimension $n-m$. Let π be the projection associated with the normal bundle $\overset{\perp}{T}(M)$. Let q be a point on the normal bundle—we can represent q by (x, ξ), where $\pi(q) = x$, and ξ is a normal vector to M at x. Now the tangent space to $\overset{\perp}{T}(M)$ at q has a canonically determined subspace, namely the space of "vertical" vectors which are tangent to the fibre through q. By parallel translation to the origin of this subspace, we can identify any vertical tangent vector at q with a vector in the vector space which is the fibre space at q. Using the euclidean metric on this space, we can certainly define an inner product of two "vertical" vectors. We now use the Riemannian structure on N to define a complementary subspace of the tangent space to $\overset{\perp}{T}(M)$ at q. We can choose a normal coordinate neighbourhood U of M centred at $\pi(q)$, so that any point of U can be joined to $\pi(q)$ by one and only one geodesic lying in U. Let r be a point of U, and let $\eta(r)$ be the uniquely determined normal vector at r obtained from ξ at $\pi(q)$ by parallel transport along the unique geodesic joining $\pi(q)$ to r with respect to the normal connexion. This determines a field of normal vectors over U, which is described by an m-dimensional section of $\overset{\perp}{T}(M)$ passing through q. The tangent space at q to this cross-section determines an m-dimensional subspace of the tangent space to $\overset{\perp}{T}(M)$ at q. We call the vectors of this subspace "horizontal", because these vectors together with the vertical vectors span the whole tangent space at q. We define the inner product of a horizontal vector and a vertical vector to be zero. All that remains is to define the inner product of two horizontal vectors. This is easily done because π_* gives a natural isomorphism between the space of horizontal vectors at q and the tangent space to M at $\pi(q)$, and we can therefore make use of the induced Riemannian metric on M. In this way we have given to $\overset{\perp}{T}(M)$ a canonical Riemannian metric.

Now the bundle of *unit* normal vectors $\overset{\perp}{T}{}^1(M)$ is a submanifold of $\overset{\perp}{T}(M)$, and the inclusion map induces on $\overset{\perp}{T}{}^1(M)$ a canonically determined Riemannian structure. In fact, this structure on $\overset{\perp}{T}{}^1(M)$ will play an important rôle in our theory.

It may be noted that because the normal connexion preserves the length of normal vectors, we could have carried out the construction directly on the unit normal bundle itself. However, the author finds it easier to deal with vector bundles rather than sphere bundles and hence we have given the rather longer method.

3.2 LIPSCHITZ-KILLING CURVATURE

Let M^n be an n-dimensional, compact, orientable C^∞-manifold and let $f: M^n \to E^{n+N}$ be a C^∞-map of M^n into euclidean space of dimension $n+N$ such that f_* is injective, that is, the Jacobian matrix of the map f when expressed in terms of local coordinates has rank n at each point $m \in M^n$. We recall that under these conditions $f: M^n \to E^{n+N}$ is called an *immersion*. If, in addition, f is injective, $f: M^n \to E^{n+N}$ is called an *imbedding*. That such maps always exist is guaranteed by Whitney's imbedding theorem which states: "*If M^n is a compact differentiable manifold, there is an injective mapping $f: M^n \to E^{n+N}$ where E^{n+N} is a euclidean space such that $f(M^n)$ is a submanifold of E^{n+N}.*"

In this chapter we show how to associate with each point of an immersed manifold and to each normal direction at that point, a real number which measures the Lipschitz-Killing curvature. The integral of the modulus of this curvature taken over the unit normal bundle gives the total absolute curvature of the immersion. The main object of this investigation is to relate the properties of the total absolute curvature to the topological properties of the manifold.

3.2.1 Fibre bundles associated with an immersion

We assume that E^{n+N} is given a definite orientation. By a frame at a point $x \in E^{n+N}$ we mean the pair consisting of x and an ordered set of mutually perpendicular unit vectors $e_1, e_2, \ldots, e_{n+N}$ such that their orientation is consistent with that of E^{n+N}. Let $F(n, N)$ be the space of all frames $(xe_1, \ldots, e_{n+N})$ of E^{n+N}. These form a differentiable manifold of dimension $\frac{1}{2}(n+N)(n+N+1)$.

In what follows it is convenient to agree to the following range of indices: $1 \leqslant i, j, k \leqslant n$; $n+1 \leqslant r, s, t \leqslant n+N$; $1 \leqslant A, B, C \leqslant n+N$.

Since E^{n+N} is $\mathbb{R}^{n+N}$ equipped with the euclidean metric, it follows that there is uniquely determined a Riemannian connexion on E^{n+N}, with zero torsion and curvature. Locally this connexion can be represented relative to an orthonormal basis (e_A) with dual basis ω_A' by a matrix of 1-forms. Instead of considering these forms locally on E^{n+N}, we may consider them to be defined *globally* over $F(n, N)$.

In terms of fibre-bundle terminology, the forms ω'_A and ω'_{AB} are defined over the bundle space of the principal bundle of frames over E^{n+N}, and the rotation group acts on the bundle space by right translation.

These forms satisfy the Cartan equations

$$de_A = \omega'_{AB} e_B :$$

$$dp = \omega'_A e_A :$$

$$d\omega'_A = \omega'_B \wedge \omega'_{BA} :$$

$$d\omega'_{AB} = \omega'_{AC} \wedge \omega'_{CB} ,$$

and the condition of using orthonormal frames with a Riemannian metric,

$$\omega'_{AB} + \omega'_{BA} = 0 .$$

Now the mapping f may be considered as a vector-valued function $f(m)$, $m \in M^n$, and f_* can be regarded as a linear differential form with values in E^{n+N}. Because of the immersion property, it has values a linear combination of n linearly independent vectors $t_1, t_2, \ldots, t_n$. We recall that a linear combination of such vectors is a tangent vector, and a vector normal to these is a normal vector. We see that the immersion of M^n in E^{n+N} gives rise to the following vector bundles:

(1) The unit tangent bundle B_τ whose bundle space is the subset of $M^n \times E^{n+N}$ consisting of all pairs (m, τ) where $m \in M^n$ and τ is a unit tangent vector at $f(m)$.
(2) The unit normal bundle B_υ whose bundle space is the subset of $M^n \times E^{n+N}$ consisting of all pairs (m, ν) where $m \in M^n$ and ν is a unit normal vector at $f(m)$.
(3) The bundle B whose bundle space is the subset of $M^n \times F(n, N)$ consisting of $(m, f(m)e_1, e_2, \ldots, e_n, e_{n+1}, \ldots, e_{n+N})$ such that $e_1, \ldots, e_n$ and $e_{n+1}, \ldots, e_{n+N}$ are respectively tangent vectors and normal vectors at $f(m)$.

We denote the projection $B \to M^n$, $(m, f(m)e_1, e_2, \ldots, e_{n+N}) \to m$ by ψ. We define $\psi_\tau : B \to B_\tau$ by $\psi_\tau(m, f(p)e_1, e_2, \ldots, e_{n+N}) = (m, e_n)$;

$$\psi_\upsilon : B \to B_\upsilon \text{ by } \psi_\upsilon(m, f(p)e_1, e_2, \ldots, e_{n+N}) = (m, e_{n+N}) .$$

Consider the maps

$$B \xrightarrow{i} M^n \times F(n, N) \xrightarrow{\lambda} F(n, N)$$

where i is the injection map and λ is projection onto the second factor. Write $\omega_A = (\lambda i)^* \omega'_A$, $\omega_{AB} = (\lambda i)^* \omega'_{AB}$. Since the operators of d and $\wedge$ commute with $(\lambda i)^*$, we deduce that ω_A and ω_{AB} satisfy the relations

$$\omega_{AB}+\omega_{BA}=0 \;,$$
$$d\omega_A=\omega_B\wedge\omega_{BA} \;,$$
$$d\omega_{AB}=\omega_{AC}\wedge\omega_{CB} \;,$$
$$dp=\omega_A e_A \;.$$

Since for points on the immersed manifold we have $\omega_r=0$, then

$$0=d\omega_r=\omega_A\wedge\omega_{Ar}=\omega_i\wedge\omega_{ir} \;.$$

Write $\omega_{ir}=A_{rij}\omega_j$; then we have $\omega_i\wedge A_{rij}\omega_j=0$, from which

$$A_{rij}=A_{rji} \;.$$

In order to familiarise ourselves with these operators, it is instructive to consider two special cases, corresponding to curves and surfaces in E^3.

3.2.2 Special cases

Case 1 $n=1$, $N=2$: $\dim F(n,N)=6$: $\dim B_\upsilon=2$: $\dim B_\tau=1$: $\dim B=2$: $i,j=1$: $r,s=2,3$: $A,B=1,2,3$.

We have $dp=\omega_1 e_1$ so $\omega_2=\omega_3=0$.

The first structural equations give $d\omega_A+\omega_{BA}\wedge\omega_B=0$:

$$d\omega_1+\omega_{21}\wedge\omega_2+\omega_{31}\wedge\omega_3=0 \quad \text{that is, } d\omega_1=0 \;.$$
$$d\omega_2+\omega_{12}\wedge\omega_1+\omega_{32}\wedge\omega_3=0 \quad \text{that is, } \omega_{12}\wedge\omega_1=0 \;:$$
$$d\omega_3+\omega_{13}\wedge\omega_1+\omega_{23}\wedge\omega_2=0 \quad \text{that is, } \omega_{13}\wedge\omega_1=0 \;.$$

Then $\omega_{12}=A_{211}\omega_1$: $\omega_{13}=A_{311}\omega_1$. The volume element of $dp=\omega^1 e_1$ is evidently equal to $\omega^1=ds$, say, when s is arc length. Also, from $de_A=\omega_{AB}e_B$ we have

$$de_1=\omega_{12}e_2+\omega_{13}e_3 \;,$$
$$de_2=\omega_{21}e_1+\omega_{23}e_3 \;,$$
$$de_3=\omega_{31}e_1+\omega_{32}e_2 \;.$$

So far we have not made a specific choice of e_2 or e_3; we do this now so that $de_1=\omega_{13}e_3$, that is, $\omega_{12}=0$. Then with $\omega_{13}=A_{311}ds$: $A_{311}=\kappa$:

$$\omega_{23}=-\tau\, ds.$$

We get the Serret-Frenet formulae on writing $e_1=t$, $e_2=b$, $e_3=n$.

Thus the Serret-Frenet formulae are contained in the structure equations of a 1-dimensional manifold imbedded in E^3.

The second structural equations involve only 2-forms, and since there are no non-trivial 2 forms on a 1-dimensional manifold, it follows that no additional information can be obtained from these equations in the case $n=1$.

Case 2 $n=2$, $N=1$: $\dim F(n,N)=6$: $\dim B_\upsilon=2$: $\dim B_\tau=3$: $\dim B=3$: $i,j,k=1,2$: $r,s=3$: $A,B=1,2,3$.

We have $dp = \omega_1 e_1 + \omega_2 e_2$ so $\omega_3 = 0$.
The first structural equations give

$$d\omega_1 + \omega_{21} \wedge \omega_2 = 0 \ ,$$
$$d\omega_2 + \omega_{12} \wedge \omega_1 = 0 \ ,$$
$$\omega_{13} \wedge \omega_1 + \omega_{23} \wedge \omega_2 = 0 \ .$$

If we write $\omega_{13} = A_{311}\omega_1 + A_{312}\omega_2$,
$\omega_{23} = A_{321}\omega_1 + A_{322}\omega_2$,

the third equation gives $A_{312} = A_{321}$.
Moreover, from the equation $de_A = \omega_{AB} e_B$ we deduce

$$-de_3 = \omega_{13} e_1 + \omega_{23} e_2 \ ;$$

the second fundamental form II, defined by $-de_3.dp$ (corresponding to $-dN.dr$ in the case of classical differential geometry of surfaces treated by classical vector calculus (see, for example I. D. Faux and M. J. Pratt (1979), p. 111, or Willmore (1959), p. 296)), is given by

$$\omega_{13}\omega_1 + \omega_{23}\omega_2 = A_{311}(\omega_1)^2 + 2A_{312}\omega_1\omega_2 + A_{322}(\omega_2)^2 \ .$$

The first fundamental form I is just $(\omega_1)^2 + (\omega_2)^2$. The principal curvatures κ_a, κ_b are by definition the eigenvalues of II with respect to I; the Gaussian curvature which is defined as the product of the principal curvatures is given by

$$K = A_{311}A_{322} - A_{312}^2 = \det(A_{3ij}) \ .$$

It is interesting to compare this with the formula equivalent to

$$\kappa = \det(A_{311})$$

obtained in the case $n = 1$.
From the second structural equations we get

$$d\omega_{12} = \omega_{13} \wedge \omega_{32} \ ,$$
$$d\omega_{13} = \omega_{12} \wedge \omega_{23} \ ,$$
$$d\omega_{23} = \omega_{21} \wedge \omega_{13} \ .$$

A straight-forward computation shows that the last two equations are just the Mainardi-Codazzi equations. The first equation may be written

$d\omega_{12} = -\det(A_{3ij})\omega_1 \wedge \omega_2$, from which we deduce that
$d\omega_{12} = -K\omega_1 \wedge \omega_2$.

An alternative approach is to assert that $\omega_1 \wedge \omega_2$ is a basis for 2-forms on

M^2 and hence $d\omega_{12}$ must differ from $\omega_1 \wedge \omega_2$ by a multiplicative scalar. So we may define a scalar-valued function K in this way. If we change the basis from (e_i) to $(e_{i'})$ by

$$e_1 = e_{1'} \cos\theta - e_{2'} \sin\theta \ , \quad \omega_1 = \omega_{1'} \cos\theta - \omega_{2'} \sin\theta \ ,$$
$$e_2 = e_{1'} \sin\theta + e_{2'} \cos\theta \ , \quad \omega_2 = \omega_{1'} \sin\theta + \omega_{2'} \cos\theta \ ,$$

then $\omega_1 \wedge \omega_2 = \omega_{1'} \wedge \omega_{2'}$.

We also readily check that $d\omega_{12} = d\omega_{1'2'}$, from which it follows that K is an invariant, determined only by the forms ω_1, ω_2. Thus we have established the fundamental theorem of Gauss, that the curvature K is an intrinsic invariant depending only upon the metric of M^2.

We also note that the volume element of the spherical indicatrix of the unit normal e_3 is

$$\omega_{13} \wedge \omega_{23} = K\omega_1 \wedge \omega_2 \ . \tag{3.10}$$

This proves Gauss's theorem:

"Let (M^2, f) be an oriented 2-manifold in E^3, and let $\nu : M^2 \to S^2$ be the spherical map which sends each point m to the point on S^2 corresponding to the unit normal at m. Then for each m, det $\nu_* = K$."

3.2.3 General Case.

We return to the general case.

The essence of Chern's approach is to consider the bundle space B and to construct differential forms in B which are the inverse images of differential forms in M^n and in B_υ, under the maps ψ and ψ_υ respectively.

The volume element of M^n can be written

$$dV = \omega_1 \wedge \omega_2 \wedge \ldots \wedge \omega_n \ : \tag{3.11}$$

From section 3.1.3 it follows that the volume element of B_υ is $dV \wedge d\sigma_{N-1}$ where $d\sigma_{N-1}$ is a form of degree $N-1$ on B_υ such that its restriction to a fibre is the volume element of the sphere of unit normal vectors at a point $m \in M^n$. We see that $de_{n+N} = \omega_{n+NA} e_A$; the normal part of this is $\omega_{n+Nr} e_r$; so the $N-1$ form required is

$$d\sigma_{N-1} = \omega_{n+N,n+1} \wedge \omega_{n+N,n+2} \wedge \ldots \wedge \omega_{n+N,n+N-1} \ . \tag{3.12}$$

Since $de_{n+N} = \omega_{n+NA} e_A$ the volume element of the unit hypersphere in E^{n+N} is

$$d\Sigma = \omega'_{n+N1} \wedge \omega'_{n+N2} \wedge \ldots \wedge \omega'_{n+N,n+N-1} \ . \tag{3.13}$$

The pull back of this form to B_υ by the dual of the differential of the spherical map ν is given by

$$\nu^*(d\Sigma) = \omega_{n+N1} \wedge \omega_{n+N2} \wedge \ldots \wedge \omega_{n+N,n+N-1} \tag{3.14}$$

that is,

$$\nu^*(d\Sigma) = (-1)^n \det(A_{n+Nij})\omega_1 \wedge \omega_2 \wedge \ldots \wedge \omega_n \wedge \omega_{n+N,n+1} \wedge \omega_{n+N,\,n+2} \wedge \ldots \wedge \omega_{n+N,n+N-1} \tag{3.15}$$

where we have used the relation $\omega_{n+Ni} = (-1)A_{n+Nij}\omega^j$. Comparing this with the form

$$dV \wedge d\sigma_{N-1} = \omega_1 \wedge \omega_2 \wedge \ldots \wedge \omega_n \wedge \omega_{n+N,n+1} \wedge \omega_{n+N,n+2} \wedge \ldots \wedge \omega_{n+N,n+N-1} \tag{3.16}$$

we see that

$$G(m, \nu) = (-1)^n \det(A_{n+Nij}) \ . \tag{3.17}$$

The scalar $G(m, \nu)$ is called the **Lipschitz-Killing curvature** of the immersion at m.

Evidently $G(m, \nu)$ depends on ν. In order to exhibit this dependence more clearly we consider a local cross-section of M^n in B, given in a neighbourhood U of m by the functions $\tilde{e}_A : q \to \tilde{e}_A(q)$ for $q \in U$. Let $e_A(q)$ be any frame in the fibre over q; then $e_A = C_{AB}e_B(q)$ where C_{AB} is an orthogonal matrix of determinant $+1$. Moreover C_{AB} has the property $C_{ir} = 0$, $C_{ri} = 0$. Using this, we find that

$$\sum_{i,j} A_{sij}\omega_i\omega_j = \sum_{r,ij} C_{sr}\tilde{A}_{rij}\tilde{\omega}_i\tilde{\omega}_j$$

where $\tilde{A}_{rij}$ is the restriction of the function A_{rij} (defined over B) to the cross-section. Let $\nu = e_{n+N}$, and write $\nu = \Sigma\nu_r\tilde{e}_r$: then at m we have

$$G(m, \nu) = (-1)^n \det(\nu_r\tilde{A}_{rij}(m)) \ .$$

For the scalar product of ν and the vector-valued second differential d^2m on B we have

$$\begin{aligned} \nu.d^2m &= \nu.d[e_A\omega_A] \\ &= \nu_r e_r.(e_A d\omega_A + \omega_A\omega_{AB}e_B) \\ &= \nu_r d\omega_r + \nu_r\omega_A\omega_{Ar} \\ &= \nu_r\tilde{\omega}_i\tilde{\omega}_{ir} = \Sigma\nu_r\tilde{A}_{rij}\tilde{\omega}_i\tilde{\omega}_j \ . \end{aligned}$$

Due to our choice of $\nu = e_{n+N}$ we see that, as forms on B_υ we have

$$-d\nu.dm = \nu.d^2m = \Sigma\nu_r\tilde{A}_{rij}\tilde{\omega}_i\tilde{\omega}_j \ .$$

Thus $G(m, \nu)$ appears as a generalisation of the determinant of the second fundamental form of a surface. In the special case when M^n is an immersed orientable hypersurface in E^{n+1}, its orientation and that of E^{n+1} define a unit normal vector $\nu_0(m)$ at $m \in M^n$. Then any other unit normal at m is of the form $\nu(m) = \pm \nu_0(m)$, and

$$G(m, \nu(m)) = G(m, \pm \nu_0(m)) = (\pm 1)^n \, G(m) \ ,$$

Thus for n even, $G(m, \nu(m))$ is independent of the orientation of the hypersurfaces and that of the space E^{n+1}. For $n=2$, $G(m)$ reduces to the Gaussian curvature K considered in case 2.

3.3 GEOMETRICAL SIGNIFICANCE OF THE LIPSCHITZ-KILLING CURVATURE

Theorem 26. *Let $L(\nu)$ be the linear space of dimension $(n+1)$ spanned by the tangent space to $f(M^n)$ at $f(m)$ and the normal vector $\nu(m)$. Then $G(m, \nu)$ is equal to the Gauss-Kronecker curvature at m of the orthogonal projection of $f(M^n)$ into $L(\nu)$.*

Proof Take a local cross-section $\tilde{e}_A(q)$ of M^n in B in a neighbourhood of m, such that $\nu(m) = \tilde{e}_{n+N}(m)$. Write $\tilde{e}_A(m) = (\tilde{e}_A)_0$, $f(m) = f_0$. Let $f'(q)$ denote the position vector of the projection of $f(q)$ in $L(\nu)$: then $f'(q)$ is defined by

$$f'(q) - f(q) = \xi_{n+1}(\tilde{e}_{n+1})_0 + \ldots + \xi_{n+N-1}(\tilde{e}_{n+N-1})_0$$
$$f'(q) - f_0 \equiv 0 \bmod (\tilde{e}_1)_0, \ldots, (\tilde{e}_n)_0, (\tilde{e}_{n+N})_0 \, .$$

From this we get

$$\xi_{n+\lambda} = (f'(q) - f(q).(\tilde{e}_{n+\lambda})_0 = (f_0 - f(q)).(\tilde{e}_{n+\lambda})_0 \ \ 1 \leqslant \lambda \leqslant N-1 \ .$$

If m is fixed and q varies on the manifold M^n we have

$$df' = df + \sum_{\lambda} d\xi_{n+\lambda}(\tilde{e}_{n+\lambda})_0 = dx - \Sigma(df.(\tilde{e}_{n+\lambda})_0)(\tilde{e}_{n+\lambda})_0$$
$$d^2f' = d^2f - \sum_{\lambda}(d^2f(\tilde{e}_{n+\lambda})_0)(\tilde{e}_{n+\lambda})_0$$

so that

$$(\tilde{e}_{n+N})_0 d^2f' = (\tilde{e}_{n+N})_0 d^2f \ .$$

This completes the proof of theorem 26. Apart from minor changes of notation, this is precisely the proof given by Chern and Lashof (1957).

3.4 TOTAL ABSOLUTE CURVATURE

Let

$$K^*(m) = \frac{1}{c_{n+N-1}} \int |G(m, \nu)| \, d\sigma_{N-1} \qquad (3.18)$$

where the integration is carried out over the sphere of unit normal vectors at $f(m)$, and where c_{n+N-1} is the volume of the unit hypersphere in E^{n+N}. We call $K^*(m)$ the **total absolute curvature of the immersion** (M^n, f, E^{n+N}) at m.

The value

$$\tau(M,f) = \int_{M^n} K^*(m)dV \tag{3.19}$$

is called the **total absolute curvature of the immersion** (M^n, f, E^{n+N}).

As we have already seen, in the case of a 2-dimensional (orientable) closed surface immersed in E^3, the Lipschitz-Killing curvature becomes the Gaussian curvature K and (3.19) becomes

$$\tau(M^2,f) = \frac{1}{2\pi} \int_S |K| dS \ . \tag{3.20}$$

The factor 2π in the denominator may seem unexpected since c_2 is equal to 4π. However, it must be remembered that the unit normal bundle becomes a double covering of M^2 and this gives rise to the 2π.

Similarly if M^n is immersed as an (orientable) closed hypersurface in E^{n+1}, (3.19) becomes

$$\tau(M^2,f) = \frac{2}{c_n} \int_{M^n} |K| dV \tag{3.21}$$

where K is the Gauss-curvature, that is, the product of the principal curvatures of the hypersurface.

Not much seems to be known about properties of the Lipschitz-Killing curvature of immersed manifolds in the general case, no doubt due to the difficulty of computing $G(m, \nu)$. Even when an explicit formula for $G(m, \nu)$ is obtained, one then has the technically difficult problem of integrating the modulus of $G(m, \nu)$.

We now prove the satisfactory result that the total absolute curvature of the immersion (M^n, f, E^{n+N}) remains unchanged if $f(M^n)$ is considered as a submanifold of a high-dimensional euclidean space which contains E^{n+N} as a linear subspace. More specifically we prove

Theorem 27. *Let* $f: M^n \to E^{n+N}$ *be an immersion of a compact differentiable manifold* M *in* E^{n+N} *given by*

$$f: m \to f(m) = (x^1(m), x^2(m), \dots, x^{n+N}(m)), \ m \in M^n \ .$$

Let $f': M^n \to E^{n+N'}$. $(N < N')$ *be the immersion defined by*

$$f'(m) = (x^1(m), \dots, x^{n+N}(m), 0, \dots, 0).$$

Then the immersed manifolds $f(M^n)$ and $f'(M^n)$ have the same total absolute curvature.

To prove the theorem, we note that it is sufficient to deal with the case $N'=N+1$, since the general case will follow by induction on the difference $N'-N$. Let a be one of the two unit vectors in E^{n+N+1} which is perpendicular to E^{n+N}. Then any unit normal vector at $f'(m)$ can be expressed uniquely in the form

$$e'_{n+N+1}=\cos\theta e_{n+N}+\sin\theta a\ ,\quad -\tfrac{1}{2}\pi<\theta\leqslant\tfrac{1}{2}\pi\ . \tag{3.22}$$

where e_{n+N} is the unit vector in the direction of its orthogonal projection on E^{n+N}. Let

$$e'_{n+N}=\sin\theta e_{n+N}-\cos\theta a\ ,\quad e'_s=e_s\,(1\leqslant s\leqslant n+N-1)\ . \tag{3.23}$$

Define a set of 1-forms $\phi_{n+N+1,A}$ by

$$\phi_{n+N+1,A}=de'_{n+N+1}.e'_A\ . \tag{3.24}$$

From our previous description it follows that the total absolute curvature of the immersion (M^n, f, E^{n+N}) is given by

$$\tau(M^n, f, E^{n+N})=\frac{1}{c_{n+N-1}}\int_{B\nu}\omega_{n+N,1}\wedge\omega_{n+N,2}\wedge\ldots\wedge\omega_{n+N,n+N-1}. \tag{3.25}$$

Similarly we have

$$\tau(M, f', E^{n+N+1})=\frac{1}{c_{n+N}}\int_{B'\nu}\phi_{n+N+1,1}\wedge\phi_{n+N+1,2}\wedge\ldots\wedge\phi_{n+N+1,n+N} \tag{3.26}$$

where B'_ν is the unit normal bundle corresponding to the immersion f'. Now from (3.22) we have

$$\begin{aligned} de'_{n+N+1}&=\cos\theta\ de_{n+N}+\{-\sin\theta e_{n+N}+\cos\theta a\}d\theta\\ &=\cos\theta de_{n+N}-e'_{n+N}d\theta\ ,\quad \text{using (3.23).}\end{aligned}$$

Moreover,

$$de'_{n+N}.e'_{n+N}=(\cos\theta d\theta e_{n+N}+\sin\theta de_{n+N}+\sin\theta d\theta a-\cos\theta da).e'_{n+N},$$

that is,

$$0=\sin\theta\cos\theta d\theta+\sin\theta de_{n+N}.e'_{n+N}-\sin\theta\cos\theta d\theta-\cos\theta da.e'_{n+N}\ .$$

Hence

$$\begin{aligned} de_{n+N}.e'_{n+N} &= \cot\theta\, da.[\sin\theta e_{n+N} - \cos\theta a] \\ &= \cos\theta e_{n+N}.da \\ &= -\cos\theta\, de_{n+N}.a \\ &= 0. \end{aligned}$$

Thus

$$\begin{aligned} \phi_{n+N+1,\,n+N} &= de'_{n+N+1}.e'_{n+N} \\ &= -d\theta\ , \end{aligned}$$

and

$$\phi_{n+N+1,s} = de'_{n+N+1}.e'_s = \cos\theta\, de_{n+N}.e_s = \cos\theta\,\omega_{n+N,s}\ .$$

Hence, from (3.26) we have

$$\begin{aligned} \tau(M,f',E^{n+N+1}) &= \frac{c_{n+N-1}}{c_{n+N}} \int_{-\frac{1}{2}\pi}^{+\frac{1}{2}\pi} |\cos^{n+N-1}\theta|\, d\theta.\tau(M,f,E^{n+N}) \\ &= 2\frac{c_{n+N-1}}{c_{n+N}} \int_0^{\frac{1}{2}\pi} \cos^{n+N-1}\theta d\theta\cdot\tau(M,f,E^{n+N})\ . \end{aligned}$$

Using the well-known formulas

$$c_k = \frac{2[\Gamma(1/2)]^{k+1}}{\Gamma((k+1)/2)},\quad \int_0^{\frac{1}{2}\pi} \cos^k\theta d\theta = \frac{\Gamma(1/2)\Gamma((k+1)/2)}{2\Gamma((k+2)/2)}\ , \qquad (3.27)$$

it follows from the previous equation that

$$\tau(M,f',E^{n+N+1}) = \tau(M,f,E^{n+N})$$

and the theorem is proved.

3.5 THE CALCULATION OF $\tau(M^n, f, E^{n+N})$

As we have previously remarked, the direct calculation of $\tau(M^n, f, E^{n+N})$ is very complicated. Fortunately there is an efficient but indirect way of computing $\tau(M^n, f, E^{n+N})$ by applying Morse theory. We now summarise the essential results of that theory which we shall subsequently use.

For an introductory text on Morse theory we refer the reader to "Morse Theory" by J. Milnor, *Annals of Mathematical Studies*, No. 51, Princeton University Press, 1963. The philosophy underlying this subject is that it is possible to obtain topological information about a differentiable manifold M by examining the behaviour of smooth real-valued functions defined over M.

Let f be a smooth real-valued function on a manifold M of dimension n. We say that f has a **critical point** at $p \in M$ if the induced map $f_* : TM_p \to T\mathbb{R}_{f(p)}$ is zero. In terms of a local coordinate system $(x^1, \ldots, x^n)$ valid in some neighbourhood of p this implies that

$$\frac{\partial f}{\partial x^1}(p) = \frac{\partial f}{\partial x^2}(p) = \ldots = \frac{\partial f}{\partial x^n}(p) = 0 \ . \tag{3.28}$$

The real number $f(p)$ is called a **critical value** of f.

A critical point is called **non-degenerate** if and only if the matrix $(\partial^2 f/\partial x^i \partial x^j)(p)$ is non-singular. It is easily verified that this condition is independent of the particular coordinate system chosen. The non-degenerate critical point is said to be of **index** λ if the matrix has λ negative eigenvalues.

Let M be compact and let f be a smooth function on M whose only critical points are non-degenerate (and therefore isolated). Let C_λ be the number of critical points of index λ, and let R_λ denote the λth Betti number, that is, the dimension of the λth homology group of M taken over some field F. Then the Morse inequalities are

$$\begin{aligned}
&R_0 \leqslant C_0 \ , \\
&R_1 - R_0 \leqslant C_1 - C_0 \ , \\
&R_2 - R_1 + R_0 \leqslant C_2 - C_1 + C_0 \ , \\
&\ldots\ldots\ldots\ldots\ldots\ldots\ldots\ ,
\end{aligned}$$

and finally the equality

$$\sum_{\lambda=0}^{n} (-1)^\lambda R_\lambda = \sum_{\lambda=0}^{n} (-1)^\lambda C_\lambda = \chi(M) \ .$$

From the above inequalities we may deduce the so-called *weak Morse inequalities*:

$$R_\lambda \leqslant C_\lambda \ . \tag{3.29}$$

We refer the reader to Sections 1–5 of Milnor's book where a proof of the above inequalities will be found. We also quote from Section 4 of the same source the following theorem due to G. Reeb:

Let M^n be a compact manifold and let f be a smooth function on M^n with precisely two critical points, both of which are non-degenerate. Then M^n is homeomorphic to the sphere S^n.

It should be noted that the theorem states that M^n is *homeomorphic* as distinct from diffeomorphic to S^n. In fact Milnor used this theorem to show that there are manifolds homeomorphic to S^7 which are not diffeomorphic to S^7 with its usual differentiable structure.

We return now to the problem of calculating $\tau(M^n, f, E^{n+N})$. The essential step is to define a real-valued function over M whose critical points are related to the geometric configuration.

We remind the reader of the unit normal bundle B_ν associated with the immersion $f\colon M^n \to E^{n+N}$. The map $B_\nu \to S_0^{n+N-1}$ which sends a unit normal vector at $f(p)$ to a point of S_0^{n+N-1} obtained by parallel translation in E^{n+N} is called the Gauss map because this is a natural generalization of the map introduced by Gauss associated with 2-dimensional surfaces in E^3. Our previous analysis shows that the total absolute curvature is the "average number of times" that S_0^{n+N-1} is covered by the Gauss map.

Now let $\nu_0 \in S_0^{n+N-1}$ and denote by ϕ the mapping which sends $p \in M$ to the scalar product of ν_0 and $f(p)$, the latter being regarded as a vector in E^{n+N}. The critical points of this real-valued function occur where $d\phi = 0$, that is, $\nu_0 . df(p) = 0$, that is, the critical points of ϕ are precisely those points on M where there is a unit normal vector parallel to ν_0. The immediate problem is to decide whether these critical points are degenerate or non-degenerate.

From our examination of the scalar product of ν_0 and the vector-valued second differential in 3.2, we see that these critical points are degenerate precisely when those points have zero Lipschitz-Killing curvature. Moreover these critical points are precisely those points where the Gauss-map does not have maximal rank. However, the theorem of Sard shows that the measure of the image of these points in S_0^{n+N-1} is zero. Thus, in computing the total absolute curvature we can ignore these degenerate points. In other words we can ignore those ν_0 such that $\nu_0 . f(p)$ has degenerate critical points, since these constitute a set of measure zero on S_0^{n+N-1}.

This, together with the Morse inequalities, gives

Theorem 28. *We have*

$$\tau(M, f, E^{n+N}) \geqslant \sum_{\lambda=0}^{n} C_\lambda(M) \geqslant \sum_{i=0}^{n} R_i(M) \ .$$

Corollary 29. *We have, for all immersions,* $\tau(M, f, E^{n+N}) \geqslant 2$.

In fact, corollary 29 follows without the use of the Morse inequalities by observing that on a compact manifold a real-valued function is either a constant or attains both its maximum and minimum value.

We can also prove

Theorem 30. *Let M^n be a compact orientable manifold immersed in E^{n+N}, such that $\tau(M^n, f, E^{n+N}) < 3$. Then M^n is homeomorphic to a sphere of n-dimensions.*

To prove the theorem, we note that our hypothesis implies that there is a set of positive measure on S_0^{n+N-1} such that if ν_0 belongs to this set, then the real-valued function $\nu_0 . f(p)$ has just two critical points. For if this were not the case we should have $\tau \geqslant 3$ giving a contradiction. Since by Sard's theorem, the image of the set of critical points under the Gauss-map is a set of measure zero, we can choose a "good" vector ν_0 such that $\nu_0 . f(p)$ has exactly two critical points on M^n, and moreover that ν_0 is the image of noncritical points of the Gauss-map. This implies that $G(p, \nu_0) \neq 0$ at each critical point of the function $\nu_0.f(p)$, that is, $\nu_0.d^2f$ is a quadratic differential form with non-zero determinant. Thus the function $\nu_0.f(p)$ defined on M^n has precisely two non-degenerate critical points. It follows from the theorem of Reeb that M must be homeomorphic to the sphere S^n.

Theorem 31. *Let M^n be a compact orientable manifold immersed in E^{n+N} such that $\tau(M^n, f, E^{n+N}) < 4$. Then M^n is homeomorphic either to the sphere S^n or else to one of the n-dimensional manifolds "like projective planes" in the sense of Eells and Kuiper.*

The proof of this theorem is analogous to that of theorem 30. We prove that there exists a function defined on M^n with either two or three non-degenerate critical points. Analogous to but more complicated than Reeb's work, Eells and Kuiper have classified manifolds with precisely 3 non-degenerate critical points in their paper Publ. Inst. Hautes Études Sci. **14** (1962), 181–222. They are called "manifolds like projective planes" for they include the real, complex, and quaternionic projective planes and the 16-dimensional Cayley plane. Thus theorem 31 is proved.

In the next chapter we shall consider those immersions of manifolds for which the total absolute curvature attains its infimum value.

Chapter 4

Tight Immersions

4.1 INTRODUCTION

In the previous chapter we found a lower bound for the total absolute curvature of the immersion M^n in E^{n+N}. We shall call the immersion $f\colon M^n \to E^{n+N}$ **proper** if the image $f(M^n)$ is not contained in a linear subvariety of E^{n+N}. If in addition such an f has minimal total absolute curvature, we say that f is a **tight** immersion. A natural question arises "for which manifolds M^n, which immersions f, and which integers N is the immersion $f\colon M^n \to E^{n+N}$ tight? Moreover, when such tight immersions exist what are the special properties of $f(M^n)$?"

Historically the first results were due to W. Fenchel [*Math. Ann. Bd.*, **101**, (1929), 238–252] and to I. Fary [Sur la courbure totale d'une courbe gauche faisant un noeud, *Bull. Soc. Math. France*, **78**, 1949, 128–138] who considered the special case $m=1$, $N=2$, that is of curves in E^3. In particular Fenchel showed that for a space curve C,

$$\tau = \frac{1}{\pi} \int_C |\kappa|\, ds \geqslant 2 \ , \tag{4.1}$$

equality holding only when C is the boundary of a convex set in E^2. Thus *a tight immersion gives a natural generalization of the important mathematical concept of convexity.*

There are many apochryphal stories associated with John Milnor but I am assured on good authority that the following is true. While Milnor was an undergraduate in his first year at Princeton, he attended lectures by Professor Tucker on differential geometry and topology. About half way through the year, towards the end of a lecture, Tucker put on the blackboard two unsolved problems in order to convince the students that the frontiers of knowledge were not far away. Within a week Milnor sought Tucker's advice about his solution to one of the problems—with typical modesty, Milnor thought that there must have been a flaw in his argument if he had solved an open problem in his year as a freshman. His solution was in fact correct, and he had proved that the total absolute curvature of a knotted curve in E^3 must be at least 4. We shall refer to Milnor's knot papers later in this chapter—Milnor (1950) and (1953).

4.2 A KEY RESULT

We now state one of the key results of our subject.

Theorem 32. *Let M^n be a compact oriented C^∞-manifold immersed in E^{n+N}, such that the total absolute curvature equals 2. Then M^n belongs to a linear subvariety E^{n+1} of dimension $n+1$, and is imbedded as a convex hypersurface in E^{n+1}. The converse is also true.*

We first prove

Lemma 33. *Let f: $M^n \to E^{n+N}$ be a tight immersion such that $\tau = 2$. Then $f(M^n)$ is immersed in a linear subvariety of dimension $n+1$ of E^{n+N}, that is $N=1$.*

Proof Assume $N \geqslant 2$, for otherwise there is nothing to prove. Our method of proof is to show that M is contained in a hyperplane of E^{n+N}, and then we argue by induction. We obtain this result by contradiction, that is, we assume that M does not lie in a hyperplane and prove that this is incompatible with the hypothesis that $\tau = 2$.

We consider a point $(p, \nu_0(p))$ of the unit normal bundle where $G(p, \nu_0) \neq 0$—such a point clearly exists since $\tau \geqslant 2$. We choose a local cross-section of M^n in the adapted frame-bundle B, that is a mapping of a neighbourhood U containing p into B such that to each point $q \in U$ there corresponds a frame $\tilde{e}_A(q)$ and such that $\tilde{e}_{n+N}(p) = \nu_o$. Any other normal vector at p can be written in the form $\Sigma \nu_r \tilde{e}_r(p)$. From Section (3.2.3) we have

$$G(p, \nu) = (-1)^n \det(\Sigma \nu_r \tilde{A}_{rij}(p)) \ . \tag{4.2}$$

We now consider the circle of unit normal vectors lying in the plane generated by $\tilde{e}_{n+N}(p)$ and $\tilde{e}_{n+N-1}(p)$. Such a vector can be written

$$\nu(\theta) = \cos\theta \tilde{e}_{n+N} + \sin\theta \tilde{e}_{n+N-1} \ , \tag{4.3}$$

from which we deduce that $G(p, \nu(\theta))$ is a polynomial in $\cos\theta$ and $\sin\theta$ and therefore an analytic function, say $g(\theta)$. Moreover, $g(\theta)$ is not identically zero, since $g(0) = G(p, \nu_o) \neq 0$.

Let H_θ be the uniquely determined tangent hyperplane at $f(p)$ which is orthogonal to $\nu(\theta)$ at $f(p)$. Since $f(M)$ does not belong to a hyperplane, it follows that there exist tangent hyperplanes H_{θ_1}, H_{θ_2} with $\theta_1 < \theta_2$ and points $q_1, q_2 \in M$ such that $f(q_1) \in H_{\theta_1}$, $f(q_2) \in H_{\theta_2}$ with $f(q_1) \notin H_{\theta_2}$ and $f(q_2) \notin H_{\theta_1}$. Since $g(\theta)$ does not vanish identically we can always choose a θ_3 such that $g(\theta_3) \neq 0$ and such that the corresponding hyperplane H_{θ_3} separates $f(q_1)$ and $f(q_2)$. The condition $g(\theta_3) \neq 0$ implies that the Gauss-map $\bar{\nu}$ is injective in a neighbourhood W of $(p, \nu(\theta_3))$ of B_υ. We can restrict W so that when $(q', \nu') \in W$, $f(q_1)$ and $f(q_2)$ are separated by the tangent hyperplane perpendicular to ν' at $f(q')$. Then the real-valued function $\nu'.f$ on M has critical points where $\nu'.df=0$, namely the point q', and the points giving the function its minimum and maximum values. Moreover, the point q' must be distinct from these two points since by construction there are points of $f(M^n)$ on either side of the tangent hyperplane at $f(q')$ perpendicular to ν'. Thus a neighbourhood of S_0^{n+N-1} is covered at least three times by the Gauss map $\bar{\nu}$. But as every point of S_0^{n+N-1} is covered at least twice, it follows that $\tau>2$. But this contradicts the hypothesis $\tau=2$. We conclude that $f(M^n)$ must lie in a hyperplane E^{n+N-1}. But from theorem 27 we know that the total absolute curvature of f: $M^n\to E^{n+N-1}$ is still 2. We apply the argument again and deduce that $f(M)$ must be immersed in E^{n+N-2} with total absolute curvature 2. Proceeding in this way, we finally conclude that $f(M^n)$ is immersed in E^{n+1}, and lemma (33) is proved.

Lemma 34. *Let f: $M^n\to E^{n+1}$ be an immersion of a compact oriented manifold M^n and let $\bar{\nu}$: $M^n\to S_0^n$ be the Gauss normal map. Let $J(p)$ be the Jacobian matrix of $\bar{\nu}$ at p, and let $U_m=\{p\in M^n|\text{rank } J(p)=n-m\}$. Then if U_m contains an open set V, its image under f is generated by m-dimensional planes.*

To prove the lemma let p be an interior point of U_m. Then since the rank of $J(p)$ is $n-m$, we may choose coordinates $t_1,\ldots,t_n$ valid in a neighbourhood of p such that if ν is the unit normal vector at $f(p)$, then $\partial\nu/\partial t_\alpha=0$ and $\partial\nu/\partial t_a$ are linearly independent. Here we have used the following convention for indices

$$1\leqslant \alpha,\beta,\gamma\leqslant m : m+1\leqslant a,b,c\leqslant n \ :$$

$$1\leqslant i,j,k\leqslant n.$$

Since ν is normal to $f(M^n)$ we have $<\nu, \partial f/\partial t_i>=0$. Differentiate with respect to t_α and use $\partial\nu/\partial t_\alpha=0$ to get $<\nu, \partial^2 f/\partial t_i\partial t_\alpha>=0$. Differentiate with respect to t_a to get $<\partial\nu/\partial t_a, \partial f/\partial t_i>+<\nu, \partial^2 f/\partial t_i\partial t_a>=0$. Putting $i=\alpha$ in the above gives

$$<\partial\nu/\partial t_a, \partial f/\partial t_\alpha>=0\ .$$

Thus the vector $\partial f/\partial t_\alpha$ is perpendicular to the $n-m$ vectors $\partial\nu/\partial t_a$ and also to ν. Hence the surfaces $t_a=\text{constant}=t_a^0$ are m-dimensional planes in E^{n+1}. Since $d\nu=\dfrac{\partial\nu}{\partial t_\alpha}\cdot dt_\alpha+\dfrac{\partial\nu}{\partial t_a}\cdot dt_a$ and $\partial\nu/\partial t_\alpha=0$, it follows that the tangent hyperplane to $f(M^n)$ remains constant along an m-dimensional generating plane. This completes the proof of the lemma.

Lemma 35. *Every boundary point of U_m which is at the same time a limit point of an m-dimensional generating plane, belongs to U_m.*

Now the proof of this lemma is rather complicated so we consider first the special case $n=2$ which will pave the way for the general case. We consider the sub-bundle B' of the adapted frame bundle B with the property that the frames e_1, e_2, e_3 are such that e_α are in the m-dimensional generating planes. To simplify matters, let us assume that $m=1$. Then

$$\alpha, \beta, \gamma = 1 : a, b, c = 2 :$$
$$1 \leqslant i, j, k \leqslant 2 .$$

From the structure equations for E^{n+1} we have, on the submanifold $f(M^n)$,

$$df = \omega_1 e_1 + \omega_2 e_2 ,$$
$$de_1 = \omega_{12} e_2 + \omega_{13} e_3 ,$$
$$de_2 = \omega_{21} e_1 + \omega_{23} e_3 ,$$
$$de_3 = \omega_{31} e_1 + \omega_{32} e_2 ,$$

where we have used the fact that $\omega_{ij} + \omega_{ji} = 0$.

Moreover we have

$$\omega_{13} = A_{11}\omega_1 + A_{12}\omega_2 ,$$
$$\omega_{23} = A_{21}\omega_1 + A_{22}\omega_2 ,$$
$$\text{with } A_{12} = A_{21} .$$

The assumption that e_1 lies in the 1-dimensional generating plane gives $\omega_{13}=0$. Thus the matrix (A_{ij}) assumes the special form

$$\begin{pmatrix} 0 & 0 \\ 0 & D \end{pmatrix}$$

where we have written $D=A_{22}$ and our assumption about the rank of $\bar{\nu}$ forces $D \neq 0$. The idea of the proof is to consider the behaviour of D as $f(p)$ moves along the generating line.

We have already shown that

$$\omega_{23} = D\omega_2 .$$

Take the exterior derivative of each side to get

$$d\omega_{23} = dD \wedge \omega_2 + D d\omega_2 .$$

From the structure equations we have

$$d\omega_{23} = \sum_k \omega_{2k} \wedge \omega_{k3} = 0 \text{ since } \omega_{13} = 0 .$$

Also

$$d\omega_2 = \sum_k \omega_k \wedge \omega_{k2} = \omega_1 \wedge \omega_{12} \ .$$

Now

$$d\omega_{13} = 0 = \sum \omega_{1k} \wedge \omega_{k3} = \omega_{12} \wedge \omega_{23} = D\omega_{12} \wedge \omega_2 \ .$$

It follows that $\omega_{12} \wedge \omega_2 = 0$, that is, we may write

$$\omega_{12} = h_{122}\omega_2 \ .$$

Then we have

$$0 = dD \wedge \omega_2 + Dh_{122}\omega_1 \wedge \omega_2 \ ,$$

that is, $0 = (dD + Dh_{122}\omega_1) \wedge \omega_2$

or $\quad dD + Dh_{122}\omega_1 = 0, \text{ mod } \omega_2 \ .$

We now return to the general case, where we wish to arrive at the following generalization of the above differential equation for D, namely,

$$dD + D\Sigma h_{\alpha aa}\omega_\alpha = 0, \text{ mod } \omega_a \ .$$

Again we refer to the structural equations of E^{n+1} in the form

$$df = \sum_i \omega_i e_i \ , \quad de_i = \sum_j \omega_{ij} e_j + \omega_{in+1} e_{n+1} \ , \quad de_{n+1} = \sum_j \omega_{n+1j} e_j \ ,$$

where $\omega_{in+1} = \Sigma A_{ij}\omega_j$, and $A_{ij} = A_{ji}$.
Our assumption on the bundle of frames B' gives

$$\omega_{\alpha n+1} = -\omega_{n+1\alpha} = 0 \ .$$

The matrix A_{ij} now takes the form

$$\begin{pmatrix} 0 & 0 \\ 0 & A_{ab} \end{pmatrix}, \text{ with } \det(A_{ab}) = D \neq 0 \ .$$

Again we study how D varies along the m-dimensional generating plane. We have

$$0 = d\omega_{\alpha n+1} = \sum_k \omega_{\alpha k} \wedge \omega_{kn+1} = \Sigma\omega_{\alpha\beta} \wedge \omega_{\beta n+1} + \Sigma\omega_{\alpha a} \wedge \omega_{an+1} \ .$$

As before $\omega_{\beta\, n+1} = 0 \ , \quad \omega_{a\, n+1} = \sum_b A_{ab}\omega_b \ ,$

so that $\sum_{a,b} A_{ab}\omega_{\alpha a} \wedge \omega_b = 0$ or $\sum_a A_{ab}\omega_{\alpha a} \wedge \prod_c \omega_c = 0$.

Since $\det(A'_{ab}) \neq 0$ we get $\omega_{\alpha a} \wedge \prod_c \omega_c = 0$.

Thus we can write

$$\omega_{\alpha a} = \sum_b h_{\alpha ab}\omega_b \ .$$

From the properties of a determinant we have

$$\prod_a \omega_{an+1} = D \prod_c \omega_c \ .$$

Take the exterior derivative of each side to get

$$\sum_a (-1)^{a-m-1}\omega_{m+1n+1} \wedge \ldots \wedge d\omega_{an+1} \wedge \ldots \wedge \omega_{nn+1}$$
$$= dD_\alpha \prod_c \omega_c + D\,(\sum_a (-1)^{a-m-1}\omega_{m+1} \wedge \ldots \wedge d\omega_a \wedge \ldots \wedge \omega_n.)$$

We substitute in the above formula for $d\omega_{an+1}$ and $d\omega_a$ from the structural equations

$$d\omega_{an+1} = \sum_k \omega_{ak} \wedge \omega_{kn+1} = \sum_b \omega_{ab} \wedge \omega_{bn+1}$$
$$d\omega_a = \sum_k \omega_k \wedge \omega_{ka} = \sum_{\alpha,b} h_{\alpha ab}\omega_\alpha \wedge \omega_b + \sum \omega_b \wedge \omega_{ba} \ .$$

We find

$$0 = dD \wedge \prod_c \omega_c + D\,(\sum_{a,\alpha} h_{\alpha aa}\omega_\alpha \wedge \prod_c \omega_c),$$

$$\text{that is, } dD + D \sum_{a,\alpha} h_{\alpha aa}\omega_\alpha = 0, \text{ modulo } \omega_c \ ,$$

which is the required relation.

Let $p \in M^n$ be a boundary point of U_m such that $f(p)$ is a limit point of the generating m-dimensional plane P. We choose a neighbourhood W of p, in which f is injective and we suppose that $f^{-1}(P) \subset W$. Let $\tilde{e}_1(q), \ldots, \tilde{e}_{n+1}(q)$, $q \in W$ be a local cross-section of W into B such that for $q \in f^{-1}(P)$, $\tilde{e}_1(q), \ldots, \tilde{e}_m(q)$ span P. Let $\tilde{\omega}_i$, $\tilde{\omega}_{ij}$ be the restrictions of ω_i, ω_{ij} to this cross-section. Then, since the $\tilde{\omega}_i$ are linearly independent we have

$$\tilde{\omega}_{\alpha a} = \sum_k \tilde{h}_{\alpha ak}\tilde{\omega}_k \ .$$

Provided that $q \in f^{-1}(P)$, the coefficients $\tilde{h}_{\alpha ab}(q)$ are none other than the functions $h_{\alpha ab}$. Let γ be a curve in $f^{-1}(P)$ which contains the point p. Then along γ we have

$$dD + D\ (\sum_{a,\alpha} h_{\alpha aa}\omega_\alpha) = 0\ .$$

Integration of this equation gives

$$D(q) = D_o \exp\ (-\int \sum_{\alpha, a} h_{\alpha aa}\omega_\alpha)\ ,$$

where $q \in \gamma - p_0$ and where $D_0 \neq 0$ is the value of D at some fixed point of γ. Since $D(q)$ is a continuous function and $h_{\alpha aa}$ is bounded it follows that $D(p) \neq 0$. Thus p belongs to U_m and the lemma is proved.

We now complete the proof of Theorem 32. Let h be a hyperplane of E^{n+1}. We can specify h as the tangent plane at the point Q on a sphere centre O and radius r, and thus it makes sense to talk about neighbouring hyperplanes in E^{n+1}. A tangent hyperplane to $f(M^n)$ is said to be of *rank m*, if it is tangent to $f(M^n)$ at a point p of $f(U_m)$ but not at a point of $f(U_l)$ where $l < m$. We first note that a tangent hyperplane of rank 0 does not separate the points of $f(M^n)$. This follows using the same type of argument in the proof of Lemma 33 for otherwise there would be a neighbourhood of S_0^n whose points are covered at least three times by the normal map $\bar{\nu}$, contradicting that the total curvature is precisely 2.

We next show that in every neighbourhood in the space H of hyperplanes of a tangent hyperplane π of $f(M^n)$ there is a tangent hyperplane of rank zero. Let $f(p)$ be a point of contact of the tangent hyperplane π of $f(M^n)$, and let W be a neighbourhood of π in the space of hyperplanes. Either there is a neighbourhood of p in M which belongs completely to U_m, or else there are points of U_l, $l < m$, in every neighbourhood of p. In either case we can find a point p_1 such that the tangent hyperplane π_1 to M at $f(p_1)$ belongs to W and such that p_1 has a neighbourhood in M which belongs completely to U_l, $l \leq m$. We have already seen that the image under f of this neighbourhood of p_1 is generated by l-dimensional planes. The key remark is that the *tangent hyperplane to f(M) along the generating l-dimensional plane through f(p₁) is precisely* π_1. Let $f(p_2)$, $p_2 \in M$, be a boundary point of this l-plane. Then, by lemma 34, p_2 belongs to U_l and is not an interior point of U_l. Thus, in every neighbourhood of p_2 there is an open set whose points are in U_k for $k < l$, and which contains a point p_3 such that the tangent hyperplane at $f(p_3)$ is in W.

Each time the above process is applied, the rank of the tangent hyperplane is effectively reduced by one. Since we start off with a finite rank m, we end up by seeing that W contains a tangent hyperplane of $f(M)$ of rank zero.

Thus we have proved that every neighbourhood of π in H contains a tangent hyperplane such that $f(M)$ lies on one side of it. Thus the same must be true of π itself. Moreover, if ν_0 is any point of S_0^n, the real-valued function $<\nu, f(p)>$ attains a maximum and a minimum on M which must be distinct, because M cannot be immersed in a hyperplane of E^{n+1}. Thus the intersection of all the half-spaces of E^{n+1} bounded by a tangent hyperplane of $f(M)$ and containing points of $f(M^n)$ is a closed convex set with a non-empty interior and with $f(M)$ on the boundary.

Since f is an immersion, it follows that f_*, the induced homomorphism of tangent spaces, is injective, and f is a local homeomorphism of M into the boundary of a convex set. Since $f(M)$ is both open and closed on the boundary, it follows that f maps M *onto* the boundary. However, the boundary of a convex set is homeomorphic to a sphere S^n, and we know that M is a covering of S^n under the map f. Thus, if $n \geqslant 2$, f must be a homeomorphism. Thus the immersion is an *imbedding*. If $n=1$, we have already proved that $f(M^1)$ must be the boundary of a convex subset of E^2, and again f must be an imbedding.

Conversely, let $f(M^n)$ be a convex hypersurface of E^{n+1}. Then, modulo a set of measure zero, by Sard's theorem every point of the unit sphere S_0 is covered exactly twice by the Gauss-map $\bar{\nu}$. Thus the total absolute curvature is 2.

This completes the proof of theorem 32.

4.2.1 Special cases

As special case of theorem 32 we have the theorem of Fenchel, which states that if a space curve c is imbedded in E^3 so that

$$\frac{1}{\pi}\int|\kappa|\,ds=2\ , \tag{4.4}$$

then c is a plane convex curve. But, of course, this result can be more quickly proved by direct methods. We give below a more direct proof of the result for surfaces immersed in E^3, namely,

$$\frac{1}{2\pi}\int_M |K|\,dS \geqslant 4-\chi(M)\ . \tag{4.5}$$

Let M be a compact closed C^∞-surface with Euler-Poincaré characteristic $\chi(M)$, immersed by mapping f into euclidean three space E^3. Then from the theorem of Gauss-Bonnet, which is also valid for non-orientable surfaces by considering the orientable double cover, we have

$$\frac{1}{2\pi}\int_M K\,dS=\chi(M)\ . \tag{4.6}$$

We split up the above integral into a contribution from the set of points for which $K>0$ and $K<0$ respectively. Hence

$$\chi(M)=\frac{1}{2\pi}\int_{K>0} K\,dS+\frac{1}{2\pi}\int_{K<0} K\,dS\ , \tag{4.7}$$

$$\tau(M,f)=\frac{1}{2\pi}\int_{K>0} KdS-\frac{1}{2\pi}\int_{K<0} KdS\ . \tag{4.8}$$

Hence

$$\tau(M,f)+\chi(M)=\frac{1}{\pi}\int_{K>0} KdS\ . \tag{4.9}$$

Now the Gauss map $M \to S^2$ is surjective—in fact, if we restrict the map to the set $K>0$, it is easy to see that the map remains surjective. Hence

$$\frac{1}{\pi}\int_{K>0} KdS \geqslant 4\ , \tag{4.10}$$

and hence

$$\tau(M,f)\geqslant 4-\chi(M)\ , \tag{4.11}$$

as required.

4.3 NECESSARY CONDITIONS FOR TIGHT IMMERSIONS

We have already seen in chapter 3 that the total absolute curvature $\tau(M^n,f)$ satisfies the inequality

$$\tau(M^n,f)\geqslant \beta(M^n)\ , \tag{4.12}$$

where $\beta(M^n)=\sum_{i=0}^{n}\beta_i$ is the sum of the Betti numbers. However, we did not specify the coefficient field relative to which the Betti numbers were to be calculated. We now show that if an immersion exists such that $\tau(M^n,f)$ attains the value $\beta(M^n)$ with respect to the coefficient field of real numbers, then the manifold M^n has zero torsion. This obviously imposes a necessary condition on the manifold M^n.

Let $\beta_i(M^n, F)$ denote the ith Betti number of M^n with respect to the coefficient field F, and let $\beta(M^n, F)$ be the sum of the numbers $\beta_i(M^n, F)$. Let $\mathbb{R}$ denote the real number field and $\mathbb{Z}_p$ the field of integers modulo prime number p. Then a standard result from algebraic topology gives

$$\beta_i(M^n, \mathbb{R})\leqslant \beta_i(M^n, \mathbb{Z}_p)\ ,\quad 0\leqslant i\leqslant n\ . \tag{4.13}$$

Suppose we assume that $\tau(M^n,f)=\beta(M^n, \mathbb{R})$. From theorem 28 we have

$$\tau(M^n,f)\geqslant \beta(M^n, \mathbb{Z}_p), \text{ which implies } \beta(M^n, \mathbb{R})\geqslant \beta(M^n, \mathbb{Z}_p)\ .$$

We thus have $\beta_i(M^n, \mathbb{R})=\beta_i(M^n, \mathbb{Z}_p)$ from which it follows that M^n has no torsion, as required.

We now refer to another interesting example of a differentiable manifold which can never be immersed in any euclidean space E^{n+N}, no matter how large N may be, so that the total absolute curvature attains its infimum value. Let M be the sphere S^7 with an exotic differentiable structure, that is one of the 28 classes of possible structures discovered by J. Milnor, which is different from the standard one (Milnor (1956)). Let us assume the S^7 is immersed with total absolute curvature equal to $\beta(S^7, F)$ which is equal to 2. It follows from lemma 33 that the image of S^7 is contained in E^8. But it is known that any exotic structure on S^7 cannot be induced from the euclidean structure from E^8—hence we have a contradiction, and the claim is established.

Perhaps the reader will believe at first that tight immersions of manifolds usually exist, and that one has to resort to exotic spheres before finding an exception. In fact, however, we now believe that the existence of a smooth tight immersion is a rare property. Evidence for this is provided by the following theorem of N. H. Kuiper (1980), which we state without proof.

Theorem 36. *Let M^n be an n-dimensional C^∞-manifold which is closed and has trivial homotopy groups π_r, $r=1, 2, \ldots, (p-1)$ where $n=2p>2$. Let f be a tight immersion of M^n into E^{n+N}. Then either $N \geqslant 3$, M^n contains a p-cycle with self-intersection number one, and*

$n=4, \quad N \leqslant 4$;

or $\quad n=8, \quad N \leqslant 6$;
or $\quad n=16, \quad N \leqslant 10$;
or $\quad N \leqslant 2$ *and M is stably parallelisable for* $n=2p>4$ (*if f is an embedding, the case* $n=2p=4$ *is also included*).

Without understanding fully the detailed statement of the above theorem, it will be seen nevertheless that the restriction of the existence of a tight immersion imposes severe restrictions on the manifolds under consideration.

4.3.1 Upper bounds for N

In this section we shall assume that M^n is tightly immersed in E^{n+N} and we shall deduce an upper bound on the possible values of N. We shall make use of some results of Morse Theory which we now recall. Denote by Φ the class of all real-valued functions on M^n which are non-degenerate, that is, all the critical points of any function $\phi \in \Phi$ are non-degenerate.

Let $\quad \beta_k(M^n) = \inf_{\Phi} \beta_k(M^n, \phi)\ , \quad k=0, 1, \ldots, m$;

$$\beta(M^n) = \inf_{\Phi} \sum_{k=0}^{n} \beta_k(M^n, \phi)\ .$$

Since the particular ϕ which minimizes $\beta_k(M^n, \phi)$ may well be different from that which minimizes $\beta_l(M^n, \phi)$ for $l \neq k$, it follows that

$$\beta(M^n) \geqslant \sum_{k=0}^{n} \beta_k(M^n) . \tag{4.14}$$

We shall now assume that equality holds. This is certainly the case when the Morse inequalities for M^n become equalities. Recent work by Cerf makes it appear that equality holds for any connected compact manifold. Thus we claim that the results of this section apply to manifolds for which

$$\beta(M^n) = \sum_{k=0}^{n} \beta_k(M^n) . \tag{4.15}$$

If this is finally proved to hold for any manifold, so much the better!

We shall also use the result due to M. Morse [*Annals of Maths.*, **71** (1959), 352–383], which states that for any connected compact manifold X^n we have $\beta_0(X^n) = \beta_n(X^n) = 1$.

Now for the theorem and its proof. First we say what we mean by the statement that a real quadratic form ψ in n variables $\xi_1, \ldots, \xi_n$ *tolerates* the $(n+1)$-tuple of non-negative integers $\beta_* = (\beta_0, \ldots, \beta_n)$. By this we mean that if k is the index of ψ, then we must have $\beta_k \neq 0$. We say that a linear system of real quadratic forms in n variables $\xi_1, \ldots, \xi_n$ tolerates β_* if each form tolerates β_* and if at least one form of the system is positive definite. Let $g(\beta_*)$ be the maximal dimension of the linear system of quadratic forms which tolerates β_*. We define $\beta_*(M^n) = \{\beta_0(M^n), \ldots, \beta_n(M^n)\}$. Then we have

Theorem 37. *Let $f\colon M^n \to E^{n+N}$ be an immersion with minimum total absolute curvature, such that*

$$\beta(M^n) = \sum_{k=0}^{n} \beta_k(M^n) .$$

Then

$$N \leqslant g(\beta_*(M^n)) . \tag{4.16}$$

To prove the theorem, let $|f(x_p)|$ be the length of the vector $f(x)$ and let $q = f(x_0)$, $x_0 \in M^n$ be a vector for which

$$|q| = |f(x_0)| = \max_{x \in M^n} |f(x)| .$$

Since f is an immersion, it follows that the vector space V^N of all linear functions, not necessarily homogeneous, which vanish at q and on the space of the tangent vectors to $f(M^n)$ at q, has dimension N. This follows because the space of normals at q has dimension N.

Let $\phi \in V^N$. Then the composite function $\phi \circ f$ is a real-valued function on M^n

which vanishes at x_0 as does all its first partial derivatives. The nature of the critical point at x_0 is determined by the matrix of second order partial derivatives evaluated at x_0. Each matrix determines a quadratic form in local coordinates $\xi_1, \ldots, \xi_n$ in some neighbourhood of $x_0 = (0, \ldots, 0)$.

We say that two real-valued differentiable functions f and g, defined on a neighbourhood of x_0 on M^n are 0-equivalent if $f(x_0) = g(x_0)$. A set of 0-equivalent functions at x_0 is called a *0-jet*. We say that f and g are 1-equivalent if, *in addition*, the first partial derivatives of both functions have the same value at x_0. A set of 1-equivalent functions at x_0 is called a *1-jet*. Similarly, we say that f and g are 2-equivalent if, *in addition*, the second order partial derivatives of f and g have the same value at x_0. An equivalence class is called a 2-jet at x_0. Clearly the set of 2-jets at x_0 whose 0-jets and 1-jets vanish form a vector space W of dimension $\frac{1}{2}n(n+1)$, this being the dimension of the space of symmetric $n \times n$ matrices. Such a 2-jet can be identified with a homogeneous quadratic form in $\xi_1, \ldots, \xi_n$.

Let π be the map which sends the function $\phi \in V^N$ to the element of W which is the 2-jet determined by ϕ. Thus $\pi : V^N \to W$. Clearly π is a homomorphism. We assert that π is a *monomorphism*—but we postpone the proof of this claim for the moment. Granting this claim, it follows that $\pi(V^N)$ is a subspace of W of dimension N. This subspace contains a positive definite 2-jet and it contains only quadratic foıms which tolerate $\beta_*(M^n)$. Thus we have

$$N \leqslant g(\beta_*(M^n)) \ .$$

We still have to justify the above assertion which we do in the form of

Lemma 38. *π: $V^N \to W^{\frac{1}{2}n(n+1)}$ is a monomorphism.*

We give a proof by contradiction. Assume that π is not a monomorphism, so that ker $\pi \neq 0$. Then there must exist a non-zero element $\psi \in V^N$ such that $\pi(\psi) = 0$. The function $\phi = q.(q-y)$, giving the scalar product in E^{n+N} of the field vector q with the vector $(q-y)$ for $y \in E^{n+N}$, clearly belongs to V^N. Moreover $\pi(\phi)$ is positive definite. Now for any real number λ we have

$$\pi(\phi + \lambda\psi) = \pi(\phi) + \lambda\pi(\psi) = \pi(\phi) \ .$$

Since $f(M^n)$ is not contained in a hyperplane of E^{n+N}, it follows that there exists a value of λ such that the hyperplane $\phi + \lambda\psi = 0$ divides $f(M^n)$ into three non-empty parts according as $\phi + \lambda\psi > 0$, $= 0$ or < 0. With this value of λ we know that the function $(\phi + \lambda\psi) \circ f$ possesses a critical point at x_0 of index n, but the value of this point is not the maximum value over M^n. Thus there exists a linear function $\chi \in \Phi$, sufficiently near to the function $\phi + \lambda\psi$, whose composition with f has two non-degenerate critical points of index n. However, this contradicts the condition $\beta_o(M^n) = \beta_n(M^n) = 1$. Thus π must be a monomorphism, and the lemma and hence the theorem is completely proved.

4.3.2 Examples

(1) Suppose that M^n is a differentiable manifold with no special properties. Since $\pi : V^N \to W^{\frac{1}{2}n(n+1)}$ is a monomorphism we must have

$$N \leqslant \tfrac{1}{2}n(n+1) .$$

(2) Suppose that M^n is the sphere S^n: then $\beta_*(S^n)=(1, 0, 0, \ldots, 0, 1)$. All the non-singular quadratic forms must be definite so that $g=1$, and hence $N=1$. This confirms a result already obtained.

(3) Suppose M^n is the 4-dimensional complex projective plane: then

$$\beta_*(P^2(\mathbb{C}))=(1, 0, 1, 0, 1) .$$

It is not difficult to verify that a matrix representation of suitable quadratic forms is

$$\begin{bmatrix} \lambda & 0 & x_1 & -x_2 \\ 0 & \lambda & x_2 & x_1 \\ x_1 & x_2 & \mu & 0 \\ -x_2 & x_1 & 0 & \mu \end{bmatrix} \quad \text{where } \lambda, \mu, x_1, x_2 \text{ are real numbers.}$$

The dimension g is equal to 4, so $N \leqslant 4$.

(4) Suppose M^n is the 8-dimensional quaternionic projective plane $P^2(\mathbb{H})$: then

$$\beta_*(P^2(\mathbb{H}))=(1, 0, 0, 0, 1, 0, 0, 0, 1) .$$

We will see below that $g=6$ so that $N \leqslant 6$.

(5) Suppose M^n is the 16-dimensional Cayley plane of octaves: then

$$\beta_*(P(\mathcal{O}))=(1, 0, 0, 0, 0, 0, 0, 0, 1, 0, 0, 0, 0, 0, 0, 0, 1) .$$

We will see below that $g=10$ so that $N \leqslant 10$.

In cases 3 and 4 and more generally in the case when $n=2m$, $\beta_0=\beta_n=1$, $\beta_m \neq 0$, $\beta_i=0$ for $i \neq 0$, m, $2m$ we see that a matrix representation of the forms of the system is given by symmetric matrices of the form

$$\begin{bmatrix} \lambda & B \\ {}^tB & \mu \end{bmatrix}$$

where the four symbols are $m \times m$ matrices. Here λ and μ are scalars, that is, multiples of the unit matrix, while B is some scalar times some orthogonal matrix, and ${}^tB$ is the transpose of B. Evidently case 2 is a special case of this. These systems of quadratic forms have been studied by Hurwitz [Hurwitz, A., *Math. Werke II*, p. 641; or *Math. Ann.*, **88** (1923), 1–25]. In the special case $m=2$, $m=4$ or $m=8$ the matrix B is known to have the following form:

$$\overbrace{\left[\begin{array}{cc}
\overbrace{\left[\begin{array}{cc}
\overbrace{\begin{bmatrix} x_1 & -x_2 \\ x_2 & x_1 \end{bmatrix}}^{2} & \begin{matrix} -x_3 & -x_4 \\ -x_4 & x_3 \end{matrix} \\
\begin{matrix} x_3 & x_4 \\ x_4 & -x_3 \end{matrix} & \begin{matrix} x_1 & -x_2 \\ x_2 & x_1 \end{matrix}
\end{array}\right]}^{4} &
\begin{matrix} -x_5 & -x_6 & -x_7 & -x^8 \\ -x_6 & x_5 & -x_8 & x_7 \\ -x_7 & x_8 & x_5 & -x_6 \\ x_8 & x_7 & -x_6 & -x_5 \end{matrix} \\
\begin{matrix} x_5 & x_6 & x_7 & -x_8 \\ x_6 & -x_5 & -x_8 & -x_7 \\ x_7 & x_8 & -x_5 & x_6 \\ x_8 & -x_7 & x_6 & x_5 \end{matrix} &
\begin{matrix} x_1 & -x_2 & -x_3 & x_4 \\ x_2 & x_1 & x_4 & x_3 \\ x_3 & -x_4 & x_1 & -x_2 \\ -x_4 & -x_3 & x_2 & x_1 \end{matrix}
\end{array}\right]}^{8}.$$

We have that $g(\beta^*)=2+2=4$, $4+2=6$, $8+2=10$ according as $m=2$, 4 or 8 respectively.

Thus when $m=2$, $N\leqslant 4$; when $m=4$, $N\leqslant 6$; and when $m=8$, $N\leqslant 10$.

Referring back to case 1, we note that when $n=2$, $N\leqslant 3$. Thus the highest dimension of euclidean space into which a surface can be tightly immersed cannot exceed 5. The question arises whether any surface can be tightly immersed in E^5 and, if so, what is the nature of the surface and the immersion. Before dealing with this we must make a diversion about non-orientable manifolds.

4.4 NON-ORIENTABLE MANIFOLDS

Although we have restricted manifolds so far to be orientable, our theory of total absolute curvature easily extends to non-orientable manifolds. It is well-known that any non-orientable manifold M possesses a double covering $\tilde{M}$ which is orientable. We *define* the total absolute curvature for M to be *one half* of the absolute value of the Lipschitz-Killing curvature when integrated over $\tilde{M}$.

As an example, consider the real projective plane $P^2(\mathbb{R})$, whose points are 1-dimensional vector subspaces of $\mathbb{R}^3$. In terms of homogeneous coordinates a point of $P^2(\mathbb{R})$ is represented by the triad of real numbers (x, y, z), not all zero, with the proviso that the same point is equally well represented by the triad $(\lambda x, \lambda y, \lambda z)$ where λ is a non-zero real number. A 1-dimensional vector subspace is represented by a line through the origin O and this meets the unit sphere in two antipodal points. The sphere S^2 is thus a double covering of the non-orientable real projective plane. If $M^2 \equiv P^2(\mathbb{R})$ is immersed in E^{2+N}, the total absolute curvature is given by

$$\tau(M^2)=\int_{M^2} \frac{|KdS|}{2\pi}$$

which makes sense whether M^2 is orientable or not.

4.5 EXISTENCE AND NON-EXISTENCE OF TIGHT IMMERSIONS

In his paper "On convex maps", *Nieuv Archief voor Wiskunde* (3), X, (1962) 147–164, N. H. Kuiper proved a most remarkable result, namely:

Theorem 39. *Let $f\colon M^2 \to E^5$ be a tight immersion of a closed surface M^2. Then M^2 must be the real-projective plane $P^2(\mathbb{R})$, f is an imbedding, and $f(M^2)$ is a real Veronese surface.*

Let $(y_1, y_2, y_3, y_4, y_5)$ be coordinates in E^5, and consider the immersion f of S^2 into E^5 given by

$$y_1 = x_0^2,\ y_2 = x_0x_1,\ y_3 = x_1^2 + x_0x_2,\ y_4 = x_1x_2,\ y_5 = x_2^2$$

where $x_0^2 + x_1^2 + x_2^2 = 1$.

From the nature of the coordinate representation of f, it is clear that the points (x_0, x_1, x_2), $(-x_0, -x_1, -x_2)$ map to the same point in E^5. We can therefore interpret f as an immersion of $P^2(\mathbb{R})$ into E^5. We leave it as an exercise to the reader to check that f is an imbedding by proving that $f(x_0, x_1, x_2) = f(x_0', x_1', x_2')$ implies that

$$(x_0', x_1', x_2') = \pm (x_0, x_1, x_2)\ .$$

Moreover the restriction of the linear function

$$L: l_1y_1 + l_2y_2 + l_3y_3 + l_4y_4 + l_5y_5$$

to $f(P^2(\mathbb{R}))$ gives

$$(L\circ f): l_1x_0^2 + l_2x_0x_1 + l_3(x_1^2 + x_0x_2) + l_4x_1x_2 + l_5y_2^2$$

where $x_0^2 + x_1^2 + x_2^2 - 1 = 0$.

A simple calculation using Lagrange's undetermined multipliers shows that for almost all linear functions L, the composite function $L\circ f$ has precisely 3 non-degenerate critical points. Thus the unit sphere is covered 3 times by the Gauss map, and therefore the immersion is tight.

Of course, we have merely verified that f does give a tight immersion of $P^2(\mathbb{R})$ in E^5. The remarkable result of Kuiper is that this is the *only* surface which can be tightly immersed in E^5, and, moreover, that its image is the Veronese surface given by the special choice of f above. In fact Kuiper's theorem has been generalized to n-dimensional manifolds tightly immersed in E^{n+N} when $N = \frac{1}{2}n(n+1)$. J. Little and W. Pohl have shown [On tight immersions of maximal codimension, *Invent. Math.*, **13** (1971), 179–204] that the only manifold which can be so immersed is the real projective n-space and again the immersion is an imbedding onto an n-dimensional Veronese manifold.

4.5.1 Tight imbeddings of $P^2(\mathbb{R})$ into E^4

We now obtain a tight imbedding of $P^2(\mathbb{R})$ into E^4. As before, let (x_0, x_1, x_2) be homogeneous coordinates for $P^2(\mathbb{R})$ and let $z_1, z_2, \ldots, z_6$ be coordinates for E^6. We define $f_1 : P^2(\mathbb{R}) \to E^6$ by

$$z = (z_1, z_2, z_3, z_4, z_5, z_6) = \left(1 \Big/ \sum_{i=0}^{2} x_i^2\right)(x_1^2, x_2^2, x_3^2, \sqrt{2}x_2x_3, \sqrt{2}x_3x_1, \sqrt{2}x_1x_2)\,.$$

Clearly $f_1(P^2(\mathbb{R})$ is contained in the sphere $S^4 : E^5 \cap S^5$ where

E^5 has equation $z_1 + z_2 + z_3 = 1$;

S^5 has equation $z_1^2 + z_2^2 + \ldots + z_6^2 = 1$.

The point C of coordinates $(\frac{1}{2}, \frac{1}{2}, 0, \frac{1}{2}, \frac{1}{2}, 0)$ satisfies both these equations and is therefore a point on S^4. We use stereographic projection π of S^4 from C into the E^4 with equations

$$z_1 + z_2 + z_3 = 1\ , \quad \text{and} \quad z_1 + z_2 + z_4 + z_5 = 0\ ,$$

which is parallel to the tangent plane to S^4 at C. The composition $f = \pi \circ f_1 :$ $P^2(\mathbb{R}) \to E^4$ is an imbedding whose image is not contained in a hyperplane E^3.

We now prove that f is tight. We have chosen E^4 in such a way that any linear function in E^4 is the restriction to E^4 of some function on E^6 of the form

$$\phi(z) = \frac{a_1(z_1 - \frac{1}{2}) + a_2(z_2 - \frac{1}{2}) + a_3z_3 + a_4(z_4 - \frac{1}{2}) + a_5(z_5 - \frac{1}{2}) + a_6z_6}{z_1 + z_2 + z_4 + z_5 - 2}\,.$$

We readily see that ϕ remains constant on any line in E^6 that passes through the point C. Hence for any point x on $P^2(\mathbb{R})$, the two functions f_1 and f have the same value. Since $z_1 + z_2 + z_3 = 1$ on the image of $P^2(\mathbb{R})$ we can use this expression to replace the terms $-\frac{1}{2}$ in the numerator and the term -2 in the denominator, and obtain $\phi(z)$ as the quotient of two homogeneous linear functions in $z_1, \ldots, z_6$. Hence

$$\phi \circ f_1(x) = \phi \circ f(x) = \frac{\chi(x_0, x_1, x_2)}{\Psi(x_0, x_1, x_2)}$$

where χ and Ψ are quadratic functions in (x_0, x_1, x_2) and $\Psi(x_0, x_1, x_2) \neq 0$ unless $(x_0, x_1, x_2) = (0, 0, 0)$. If we take $\Psi(x_0, x_1, x_2) = 1$ we see, as in the previous example, almost every linear function in E^4 has three non-degenerate critical points. Since $\beta(P^2(\mathbb{R})) = 3$, it follows that f is a tight imbedding.

4.5.2 Tight immersions of surfaces

We have now exhibited tight immersions of $P^2(\mathbb{R})$ in E^5 and in E^4. However, in a remarkable paper by N. H. Kuiper, "On surfaces in euclidean three-space", *Bull. Soc. Math. de Belgique*, **12**, 1960, pp. 5–22, it is proved that a tight immersion of $P^2(\mathbb{R})$ in E^3 cannot exist. This also holds for the non-orientable surface called the Klein bottle, which can be described as follows. Let a, b be positive real numbers and let Γ be the group of transformations of E^2 to itself generated by

$$(x, y) \to (x, y+b) \ ,$$

$$(x, y) \to (x+\frac{a}{2}, -y) \ .$$

Clearly such transformations form a group. We identify points of E^2 which can be transformed by an element of Γ. The quotient surface thus obtained, denoted by E^2/Γ with the induced metric from E^2, is a flat compact non-orientable surface called the **Klein bottle** and is denoted by $K(a, b)$.

In this paper, Kuiper shows that with one exception, every other non-orientable surface can be tightly immersed in E^3. The exceptional surface is that for which the Euler characteristic is -1. Whether or not this surface can be tightly immersed in E^3 *is an open problem.*

On the other hand, every orientable surface can be tightly immersed in E^3 so that $\tau = 4 - \chi$. For example, the torus T^2 when immersed as an anchor ring clearly has the property that almost every point on the unit sphere is covered twice by the Gauss map. Thus, in this case we have

$$\frac{1}{2\pi} \int_T |K| dS = 4 \ . \tag{4.17}$$

4.6 Knotted tori

An interesting recent development due to R. Langevin and H. Rosenburg, "On Curvature Integrals and Knots", *Topology*, Vol. 15, 1976, pp. 405–416, contains a kind of surface analogue of Milnor's theorem on knotted curves.

Theorem 40. *Let T be a knotted torus T^2 in E^3. Then*

$$\frac{1}{2\pi} \int_T |K| dS \geqslant 8 \ . \tag{4.18}$$

Moreover, they show that there exists a knotted torus such that the integral can become as close as we please to 8. A very recent (unpublished) result due to Kuiper shows that the sign $\geqslant$ in the theorem can be replaced by the strictly greater than sign, that is, a torus for which the integral is equal to 8 is unknotted.

The results of Langevin and Rosenburg for surfaces of genus $g = 1$ have been generalized to surfaces of arbitrary genus by Meeks and by Morton. In particular they prove:

Theorem 41. *If a smooth embedded closed orientable surface $f\colon M^2 \to E^3$ has total absolute curvature*

$$\tau(M^2, f) < 2g + 6 = 8 - \chi(M) , \tag{4.19}$$

then f is regularly isotopic through embeddings to a standard ("*unknotted*") *embedding.*

The proof is quite complicated and will be omitted here. Kuiper has conjectured that the inequality sign in this theorem can be replaced by less than or equals, that is he believes that equality cannot be realized for knotted M^2, and as remarked above, he has proved this to be the case for knotted tori.

Another theorem of similar type is the following due to Ferus.

Theorem 42. *A smooth immersion of an exotic sphere Σ^n into E^{n+2} has total absolute curvature $\geqslant 4$.*

Again the question of whether the value 4 can be attained is still an open problem.

4.7 Kobayashi's Theorem

Let M be a (real) $2n$-dimensional analytic manifold which admits a field of endomorphisms J such that $J^2 = -$identity. We assume that M carries a Riemannian metric $<, >$ with the property that

$$<JX, JY> = <X, Y>$$

for all vector fields X, Y on M. We assume also that J satisfies an integrability condition given by

$$[X, Y] + J[JX, Y] + J[X, JY] - [JX, JY] = 0 . \tag{4.20}$$

Finally we assume that J satisfies the condition $\nabla_X J = 0$. Such a manifold M is a complex manifold with a Kähler structure—briefly a *Kähler manifold.*

It may happen that a Kähler manifold admits a group G of transformations onto itself which preserves the Kähler structure. Suppose that the group acts transitively, that is, given any two points P_0 and P of M there is an element of G which takes P_0 to P. Then M is called a homogeneous space. Suppose that the set of elements of G which leave P_0 fixed is H, called the isotropy subgroup of G. Then it can be shown that there is a one-one correspondence between points of M and cosets of G with respect to H. We assume that H is a closed subset of G. Then we can write $M = G/H$. The theorem of Kobayashi states that *every compact homogeneous Kähler manifold $M = G/H$ embeds with minimum total curvature into a euclidean space whose dimension is that of the group G.*

We refer the reader to the original research paper by Kobayashi (1967)

"Imbeddings of homogeneous spaces with minimum total curvature", *Tohoku Math. J.*, **19**, (1967) pp. 63–70 as although the result is very interesting, the detailed proof requires some technical knowledge of Lie groups beyond the scope of this book.

Chapter 5

Total Mean Curvature

5.1 AN EARLY RESULT

In this chapter we shall be concerned with the mean curvature of immersed manifolds. The first result of this subject due to the present author (1965) is

Theorem 43. *Let M^2 be a closed orientable C^∞-surface and let $f\colon M^2 \to E^3$ be a C^∞-immersion of M^2 into euclidean three-space. Let H be the mean curvature of the immersed surface. Let*

$$\tau(f, M^2) = \frac{1}{2\pi} \int_{M^2} H^2 dS \ . \tag{5.1}$$

Then $\tau(f, M^2) \geqslant 2$; moreover $\tau(f, M^2) = 2$ if and only if M^2 is embedded as the standard (round) sphere.

We give a simple *ad hoc* proof of this result. Later we shall prove a much more general theorem of which the above is a special case. It will be recalled that, in the classical local theory of surfaces in E^3, at each point on $f(M)$ there are two principal curvatures κ_1, κ_2. The product $\kappa_1\kappa_2$ is none other than the Gaussian curvature, and the mean $\frac{1}{2}(\kappa_1+\kappa_2)$ is the mean curvature H. Clearly $H^2 - K = \frac{1}{4}(\kappa_1 - \kappa_2)^2 \geqslant 0$. If we divide the surface $f(M^2)$ into regions for which $K>0$, $K=0$ and $K<0$, we have

$$\int_{M^2} H^2 dS \geqslant \int_{K>0} H^2 dS \geqslant \int_{K>0} K dS \geqslant 4\pi \ , \tag{5.2}$$

where we have again used the fact that the restriction of the Gauss map to the region of $f(M^2)$ where $K>0$ is surjective.

Thus we get immediately

$$\frac{1}{2\pi} \int_M H^2 dS \geqslant 2 \ . \tag{5.3}$$

Moreover, equality holds only if $\kappa_1 = \kappa_1$ at every point, that is, every point is an umbilic. It follows [Willmore (1959), p. 128] that $f(M^2)$ is embedded as the round sphere.

5.2 CONFORMAL INVARIANTS

We now prove an interesting result due to J. H. White [A global invariant of conformal mappings in space, *Proc. Amer. Math. Soc.*, **38**, 1973, 162–164]. It is fair to say, however, that this result was essentially known to W. Blaschke [*Vorlesungen über Differentialgeometrie, III*, Springer, Berlin, 1929]. Let $X: M^2 \to E^3$ be a smooth immersion of a compact orientable surface M^2 into euclidean 3-space. Let

$$T = \int_{M^2} H^2 dS$$

where H is the mean curvature. *Then T is invariant under conformal transformations of E^3.*

It is a standard result that a conformal transformation of E^3 can be decomposed into a product of similarity transformations and inversions. By a similarity transformation we mean euclidean motions and homotheties, that is, constant scale factor transformations. We now prove that T is invariant under inversions. We take the centre of inversion to be the origin, which we assume does not lie on the surface M^2. Let c be the radius of inversion. To the point $\boldsymbol{x}$ on the surface M^2 we associate the inverse point $\bar{\boldsymbol{x}}$ where

$$\bar{\boldsymbol{x}} = c^2 \boldsymbol{x}/r^2 \text{ where } r^2 = \langle \boldsymbol{x}, \boldsymbol{x} \rangle .$$

We have

$$d\bar{\boldsymbol{x}} = \frac{c^2}{r^2} d\boldsymbol{x} - \frac{2c^2}{r^3} dr\boldsymbol{x} ,$$

with

$$rdr = \langle \boldsymbol{x}, d\boldsymbol{x} \rangle .$$

Hence

$$\langle d\bar{\boldsymbol{x}}, d\bar{\boldsymbol{x}} \rangle = \frac{c^4}{r^4} \langle d\boldsymbol{x}, d\boldsymbol{x} \rangle - \frac{4c^4}{r^5} \langle d\boldsymbol{x}, dr\boldsymbol{x} \rangle + \frac{4c^4}{r^6} r^2 dr^2 ,$$

that is

$$\langle d\bar{\boldsymbol{x}}, d\bar{\boldsymbol{x}} \rangle = \frac{c^4}{r^4} \langle d\boldsymbol{x}, d\boldsymbol{x} \rangle , \tag{5.4}$$

since the last two terms on the right hand of the previous equation cancel. If dS and $d\bar{S}$ denote the corresponding volume elements of M^2 and $\bar{M}^2$, it follows that

$$d\bar{S}=\frac{c^4}{r^4}dS \ . \tag{5.5}$$

Now let $\boldsymbol{e}_3$ be the unit normal vector to M^2 at $\boldsymbol{x}$, and let $\bar{\boldsymbol{e}}_3$ denote the corresponding normal to $\bar{M}^2$ at $\bar{\boldsymbol{x}}$.

Then

$$\bar{\boldsymbol{e}}_3=\frac{2}{r^2}<\boldsymbol{x},\boldsymbol{e}_3>\boldsymbol{x}-\boldsymbol{e}_3 \ ,$$

since the right-hand member is a unit vector perpendicular to $d\bar{\boldsymbol{x}}$.

Thus for the second fundamental form of $\bar{M}$ we have

$$-d\bar{\boldsymbol{e}}_3\cdot \mathrm{d}\bar{\boldsymbol{x}}=\frac{c}{r^4}<d\boldsymbol{x},d\boldsymbol{e}_3>-\frac{2c^2}{r^4}<\boldsymbol{x},\boldsymbol{e}_3><d\boldsymbol{x},d\boldsymbol{x}> \ . \tag{5.6}$$

Combining (5.4) and (5.6) it follows that the principal curvatures κ_1, κ_2 of M^2 with respect to $\boldsymbol{e}_3$ satisfy

$$\bar{\kappa}_i=-\frac{r^2}{c^2}\kappa_i-\left(\frac{2r^2}{c^2}\right)<\boldsymbol{x},\boldsymbol{e}_3> \ , \quad i=1,2$$

where $\bar{\kappa}_i$ denotes the corresponding principal curvatures of $\bar{M}^2$ with respect to $\bar{\boldsymbol{e}}_3$.

Thus we get

$$(\bar{\kappa}_1+\bar{\kappa}_2)-4\bar{\kappa}_1\bar{\kappa}_2=\frac{r^4}{c^4}\{(\kappa_1+\kappa_2)^2-4\kappa_1\kappa_2\} \ ,$$

that is

$$\bar{H}^2-\bar{K}=\frac{r^4}{c^4}(H^2-K) \ , \tag{5.7}$$

where $\bar{H}$ and $\bar{K}$ are the mean and Gauss curvatures of $\bar{M}^2$.

From (5.5) and (5.7) we get

$$(\bar{H}^2-\bar{K})d\bar{S}=(H^2-K)dS \ , \tag{5.8}$$

a result already obtained by G. Thomsen [*Abh. Math. Sem. Univ., Hamburg* (1923), 31–56]. On integration we have

$$\int_{M^2}(H^2-K)dS=\int_{M^2}H^2dS-2\pi\chi(M^2)=\int_{M^2}\bar{H}^2d\bar{S}-2\pi\chi(M^2) \tag{5.9}$$

by the Gauss-Bonnet theorem. It follows that

$\int\limits_{M^2} H^2 dS$ *is a conformal invariant.*

We may note that almost the same proof given above applies when M^2 is embedded in euclidean space E^{2+p}, where $p \geqslant 2$. Instead of the uniquely determined normal vector $\boldsymbol{e}_3$, we take any p mutually orthogonal unit normal vectors on M^2, $\boldsymbol{e}_\alpha$, $\alpha = 3, \ldots, p+2$. This time we find that

$$\bar{\boldsymbol{e}}_\alpha = \frac{2}{r^2} < \boldsymbol{x}, \boldsymbol{e}_\alpha > \boldsymbol{x} - \boldsymbol{e}_\alpha$$

and

$$-d\bar{\boldsymbol{x}} \cdot d\bar{\boldsymbol{e}}_\alpha = \frac{c^2}{r^2} < d\boldsymbol{x}, d\boldsymbol{e}_\alpha > - \frac{2c^2}{r^4} < \boldsymbol{x}, \boldsymbol{e}_\alpha > < d\boldsymbol{x}, d\boldsymbol{x} > \ .$$

Let $\boldsymbol{e}$ be any unit vector in the normal space to M^2 at $\boldsymbol{x}$. Then, just as before, we find that the principal curvatures satisfy

$$(\bar{\kappa}_1(\bar{e}) + \bar{\kappa}_2(\bar{e}))^2 - 4\bar{\kappa}_1(\bar{e})\bar{\kappa}_2(\bar{e}) = \frac{r^4}{c^4}\{(\kappa_1(e) + \kappa_2(e))^2 - 4\kappa(e)\kappa(e)\} \ .$$

We now average over the spheres of normal vectors of M and M^2 to get

$$\bar{H}^2 - \bar{K} = \left(\frac{r^4}{c^2}\right)(H^2 - K) \text{ as before.}$$

It follows that $\int\limits_{M^2} H^2 dS$ *is a conformal invariant for embeddings of* M^2 *in* E^{2+p}. [B. Y. Chen, An invariant of conformal mappings, *Proc. Amer. Math. Soc.*, **40**, 1973, 563–564.]

5.3 IMBEDDINGS OF THE TORUS

In Chapter 3 we defined total absolute curvature and in Chapter 4 we considered tight immersions, that is, the immersions for which the total absolute curvature attained its infimum value At the beginning of this chapter we proved that $T \geqslant 2$, the infimum value 2 being attained only when M^2 is a sphere which is embedded as the round sphere In Willmore (1965), Note on embedded surfaces, *An. St. Univ.*, "Al. I. Cuza" Iasi Sect. Ia Mat. 11B (1965), 493–496, MR 34, # 1940, I suggested the problem of finding the value of

$$\inf_i \int\limits_{i(M)} H^2 dS$$

where i ranges over all embeddings of the compact orientable surface M^2 in E^3. It is clear that the infimum embedding is determined only up to a conformal transformation.

Example. Let us consider the standard embedding of the torus T^2 in E^3 given by

$$x=(a+b\cos u)\cos v\ ,\quad y=(a+b\cos u)\sin v\ ,\quad z=b\sin u\ .$$

Show that $E=b^2$, $F=0$, $G=(a+b\cos u)^2$; and that the coefficients of the second fundamental form are given by

$$L=b\ ,\quad M=0\ ,\quad N=(a+b\cos u)\cos v\ .$$

The mean curvature

$$H=\frac{a+2b\cos u}{2b(a+b\cos u)}\,,$$

and we have

$$T=\frac{1}{2\pi}\int_0^{2\pi}\int_0^{2\pi} H^2 b(a+b\cos u)\,du\,dv\ .$$

After some computation we find, on writing $b/a=c$, so that $0<c<1$,

$$T=\pi/2c(1-c^2)^{\frac{1}{2}}\ . \quad = \frac{\pi}{2c}\sqrt{1-c^2}$$

The minimum value of T occurs when $c=1/\sqrt{2}$. Then

$$\int_{M^2} H^2 dS=2\pi^2\ .$$

Clearly any surface obtained from this particular torus (the so-called **Clifford torus**) by inverting with respect to a centre not lying on the torus, will give the same value $2\pi^2$ to this integral.

I conjectured that $2\pi^2$ is the infimum value for *any* torus in E^3. In Chapter 6 we shall deal with this conjecture in some detail. We now prove a result which adds evidence for the validity of the conjecture.

Theorem 44. *Let M^2 be a torus embedded in E^3 as a tube of constant circular cross-section. More precisely, let the embedded surface be formed by carrying a (small) circle round a closed space curve γ (for which $\kappa\neq 0$) such that the centre moves along the curve and the plane of the circle is in the normal plane to the curve at each point. Then*

$$\int_{M^2} H^2 dS\geqslant 2\pi^2\ ,$$

equality holding if and only if the generating curve is a circle and the ratio of the radii is $1/\sqrt{2}$.

To prove the theorem by elementary means, we take the embedded surface as given by

$$\boldsymbol{R}=\boldsymbol{r}(s)+\varepsilon\boldsymbol{n}(s)\cos v+\varepsilon\boldsymbol{b}(s)\sin v\ ,$$

where $\boldsymbol{t}$, $\boldsymbol{n}$, $\boldsymbol{b}$ denote the Frenet frame of γ, ε is the radius of the generating circle and v is the angle of rotation.

A straightforward computation of the 6 fundamental coefficients gives

$$\begin{aligned}E&=(1-\kappa\varepsilon\cos v)^2+\tau^2\varepsilon^2\ ,\\ F&=\varepsilon^2\tau\ ,\\ G&=\varepsilon^2\ ,\\ L&=\tau^2\varepsilon-\kappa\cos v(1-\kappa\varepsilon\cos\gamma)\\ M&=\varepsilon\tau\ ,\\ N&=\varepsilon\ .\end{aligned}$$

Again a straightforward computation gives for the principal curvatures

$$\kappa_1=1/\varepsilon\ ,\quad \kappa_2=\kappa\cos v/(\kappa\varepsilon\cos v-1)\ .$$

The mean curvature is given by

$$H=(1-2\kappa\varepsilon\cos v)/2\varepsilon(1-\kappa\varepsilon\cos v)\ .$$

Computation now gives

$$\int_{M^2}H^2dS=\frac{\pi}{\varepsilon}\int_0^l\frac{ds}{\sqrt{(1-\kappa^2\varepsilon^2)}},$$

where l is the length of the curve γ. It is interesting to note that this expression is independent of the torsion of the curve γ. Thus

$$\int_{M^2}H^2dS=\frac{\pi}{2}\int_0^l\frac{|\kappa|dS}{|\kappa\varepsilon|\sqrt{(1-\kappa^2\varepsilon^2)}}\geqslant\pi\int_0^l|\kappa|ds\ ,$$

the last step following because, for any real number z, the expression $z\sqrt{(1-z^2)}$ takes its maximum value 1/2 when $z=1/\sqrt{2}$. However, from an early result in Chapter 3 (Fenchel's theorem) we have

$$\int_0^l|\kappa|ds\geqslant 2\pi\ .$$

Thus $\int_{M^2}H^2dS\geqslant 2\pi^2$ giving the required result.

Moreover equality takes place only when $\kappa\varepsilon=1/\sqrt{2}$, that is, when the torus is embedded as an anchor ring with generating circles with ratio $1:\sqrt{2}$.

An analogous result to theorem 44 was obtained independently by

K. Shiohama and A. Takagi, *Journ. Diff. Geometry* **4** (1970), 477–485: M.R. 43, # 2646.

5.4 MINIMAL SUBMANIFOLDS OF RIEMANNIAN MANIFOLDS

We wish to consider integrals involving the mean curvature in a more general situation, usually of an n-dimensional manifold M in a euclidean space N of dimension $n+p$ or in a sphere S of dimension $n+p$ which itself is imbedded with its usual metric in euclidean space of dimension $n+p+1$. At first we consider the general case $f\colon M \to X$, where f is a differentiable immersion of an n-dimensional manifold into a Riemannian manifold X of dimension $n+p$. We impose on M the induced metric from X.

We use the following range of indices

$$1 \leqslant i, j, k, \ldots \leqslant n : n+1 \leqslant \alpha, \beta, \gamma, \ldots \leqslant n+p \ .$$
$$1 \leqslant A, B, C, \ldots \leqslant n+p \ .$$

We choose an orthonormal frame field (e_A) over X so that

$$<e_A, e_B> = \delta_{AB} \ , \tag{5.10}$$
$$ds^2 = \Sigma \omega_A{}^2 \tag{5.11}$$

where (ω_A) is the dual frame of 1-forms.

The Riemannian connexion form ω_{AB} satisfies

$$\omega_{AB} + \omega_{BA} = 0 \ , \tag{5.12}$$
$$d\omega_A = \sum_B \omega_B \wedge \omega_{BA} \ . \tag{5.13}$$

The mapping f^* pulls back the tangent bundle TX over X to the bundle $f^*(TX)$ over M. We restrict to a neighbourhood of M and consider a frame field $l_A(m)$, $m \in M$ such that $l_i(m)$ are tangent vectors and $l_\alpha(m)$ are normal vectors at m. Let θ_A, θ_{AB} denote the forms previously denoted by ω_A, ω_{AB} relative to this frame field. Then we have

$$\theta_\alpha = 0 \ .$$

We take the exterior derivative of this equation to get $d\theta_\alpha = 0$ and hence from (5.13) we get

$$\Sigma \theta_i \wedge \theta_{i\alpha} = 0 \ . \tag{5.14}$$

We write

$$\theta_{i\alpha} = \sum_j h_{i\alpha j} \theta_j \ ; \tag{5.15}$$

from (5.14) we get

$$\Sigma h_{i\alpha j}\theta_j \wedge \theta_i = 0$$

from which, because the two-forms $\theta_j \wedge \theta_i$ are linearly independent, we deduce

$$h_{i\alpha j} = h_{j\alpha i} \ . \tag{5.16}$$

We write

$$\Theta_\alpha = \sum_{i,j} h_{i\alpha j}\theta_i\theta_j \ , \tag{5.17}$$

and denote by Θ the vector valued form

$$\Theta = \sum_\alpha \Theta_\alpha \otimes l_\alpha \ . \tag{5.18}$$

This is called the **second fundamental form of** M *in* X.

The **mean curvature vector** is defined by

$$H = \frac{1}{n} \sum_{i,\alpha} h_{i\alpha i} l_\alpha \ . \tag{5.19}$$

Clearly H is a normal vector field over M.

M is called a **minimal submanifold** if $H = 0$.

M is called **totally geodesic** if $\Theta = 0$.

If $n = 1$ these notions coincide, and minimal submanifolds of dimension 1 are precisely geodesics.

Exercise

Verify that if M is a 2-dimensional surface submanifold in E^3, our definitions of second fundamental form and mean curvature are consistent with the formulas

$$II \equiv L du^2 + 2M du dv + N dv^2 \ ,$$

$$H = \frac{EN + GL - 2FM}{2(EG - F^2)} \ ,$$

of classical differential geometry (e.g. Willmore (1959), p. 95 (1.5) and p. 98 (2.6)).

5.4.1 The variational equations

We assume now that M is compact and without boundary. Then its volume V is given by

$$V=\int_M \theta_1 \wedge \theta_2 \wedge \ldots \wedge \theta_n \ . \tag{5.20}$$

We consider a variation of M as follows: Let I be the interval $-\frac{1}{2}<t<\frac{1}{2}$. Let $F: M\times I \to X$ be a differentiable mapping such that for each $t \in I$, its restriction to $M\times t$ is an immersion and such that $F(m, 0)=f(m)$, $m \in M$. We take a frame field $e_A(m, t)$ over $M \times I$ such that for every $t \in T$, $e_i(m, t)$ are tangent vectors to $F(m \times t)$ at (m, t) and $e_\alpha(m, t)$ are normal vectors. We then have

$$\omega_i=\theta_i+a_i dt \ , \quad \omega_\alpha=a_\alpha dt \ , \quad \omega_{i\alpha}=\theta_{i\alpha}+a_{i\alpha}dt \tag{5.21}$$

where θ_i, $\theta_{i\alpha}$ are linear differential forms on M whose coefficients may vary with t. For $t=0$ these forms are those previously denoted by θ_i, $\theta_{i\alpha}$ on M.

The vector $\Sigma a_A e_A$ at $t=0$ will be called the **deformation vector**. We write the operator d on $M\times I$ as

$$d=d_M+dt\frac{\partial}{\partial t} \ .$$

We have

$$d(\omega_1 \wedge \omega_2 \wedge \ldots \wedge \omega_n)$$
$$=d\omega_1 \wedge \omega_2 \wedge \ldots \wedge \omega_n - \omega_1 \wedge d\omega_2 \wedge \ldots \wedge \omega_n$$
$$+\omega_1 \wedge \omega_2 \wedge d\omega_3 \wedge \ldots \wedge \omega_n - \ldots ,$$

that is $$d(\omega_1 \wedge \ldots \wedge \omega_n)=\Sigma \omega_\alpha \Omega_\alpha \ , \tag{5.22}$$

where

$$\Omega_\alpha=-\Sigma \omega_1 \wedge \ldots \wedge \omega_{i-1} \wedge \omega_{i\alpha} \wedge \omega_{i+1} \wedge \ldots \wedge \omega_n \ .$$

In performing the above calculation we have used equation (5.13).

Substituting from (5.21) in (5.22), the left-hand side becomes

$$d\{\theta_1 \wedge \theta_2 \wedge \ldots \wedge \theta_n + dt \wedge \sum_i (-1)^{i-1} a_i \theta_1 \wedge \ldots \wedge \theta_{i-1} \wedge \theta_{i+1} \wedge \ldots \wedge \theta_n\} \ .$$

The right-hand side becomes

$$dt \wedge \sum_\alpha a_\alpha \tilde{\Theta}_\alpha$$

where

$$\tilde{\Theta}_\alpha=-\sum_i \theta_1 \wedge \ldots \wedge \theta_{i-1} \wedge \theta_{i\alpha} \wedge \theta_{i+1} \wedge \ldots \wedge \theta_n \ .$$

Equate the terms in dt to get

$$\frac{\partial}{\partial t}(\theta_1\wedge\ldots\wedge\theta_m)=d_M\sum_i(-1)^{i-1}a_1\theta_1\wedge\ldots\wedge\theta_{i-1}\wedge\theta_{i+1}\wedge\ldots\wedge\theta_n+\sum_\alpha a_\alpha\tilde{\Theta}_\alpha\,.$$

Integrate over M and evaluating at $t=0$, we find

$$\begin{aligned}V'(0)&=\frac{\partial}{\partial t}\int_M\theta_1\wedge\theta_2\wedge\ldots\wedge\theta_n|_{t=0}\\&=\int_M\sum_\alpha a_\alpha\Theta_\alpha\ ;\end{aligned}$$

the term involving $\int_M d_M(\ldots)$ makes a zero contribution because, by Stokes' theorem, this integral is equal to $\int_{\partial M}(\ldots)$, and the boundary M is empty. Thus the condition $V'(0)=0$ is satisfied for arbitrary a_α if and only if

$$\Theta_\alpha=0\ . \tag{5.23}$$

These conditions are equivalent to

$$\sum_i\theta_1\wedge\theta_2\wedge\ldots\wedge\theta_{i-1}\wedge\sum h_{i\alpha j}\theta_j\wedge\theta_{i+1}\wedge\ldots\wedge\theta_n=0\ ,\quad\text{that is }\sum h_{i\alpha i}=0\ ,$$

that is $$H=\frac{1}{n}\sum_{i,\,\alpha}h_{i\alpha i}l_\alpha=0\ . \tag{5.24}$$

Thus we have proved that a closed minimal submanifold of a riemannian manifold is locally characterized by the property that its volume is stationary under deformations.

Example 1 Let M be a surface in euclidean 3-space E^3 given by

$$x=r\cos\phi\ ,\quad y=r\sin\phi\ ,\quad z=a\cosh^{-1}r/a$$

where a is a non-zero constant. Then M is a minimal surface, called a *catenoid.*

Example 2. Let M be a surface in E^3 given by

$$x=r\cos\phi\ ,\quad y=r\sin\phi\ ,\quad z=a\phi$$

where a is a non-zero constant. Then M is a minimal surface, called a *right helicoid.*

5.5 MINIMAL SUBMANIFOLDS IN EUCLIDEAN SPACE

Let M be a Riemannian manifold. For any differentiable function f, we define **grad** f, by the unique vector field X such that

$$g(X,Y)=df(Y)=Yf \text{ for all vector fields } Y.$$

In terms of local coordinates $(\text{grad } f)^i = g^{ij}\partial_j f$.

Let X be a vector field on M. We define the **divergence** of X, div X to be the function such that, at each point $x \in M$,

$$(\text{div } X)_x = \text{trace of the endomorphism}$$

of the tangent space $T_x(M)$ of M at x given by $V \to \nabla_V X$.

In terms of local coordinates, if $X = X^i \dfrac{\partial}{\partial x^i}$, then

$$\text{div } X = \sum_{i=1}^{n} X^i_{\ ,i} \; .$$

The Laplacian Δ on M is a mapping which sends any differentiable function f to the function $\Delta f = \text{div}(\text{grad } f)$. In terms of local coordinates

$$\Delta f = \sum_{i,j=1}^{n} g^{ij} f_{,ij} = \sum_{i,j=1}^{n} \frac{1}{\sqrt{g}} \frac{\partial}{\partial x^i}\left(g^{ij}\sqrt{g}\,\frac{\partial f}{\partial x^j}\right) .$$

We have

$$\begin{aligned}\Delta(f^2) &= g^{ij} f^2_{\ ;ij} = g^{ij} 2(f f_i)_{,j} \\ &= 2f\Delta f + 2g^{ij} f_{,i} f_{,j} \,.\end{aligned}$$

We now prove a result due to E. Hopf.

Lemma 45. *Let M be a closed Riemannian manifold. If f is a differentiable function on M such that $\Delta f \geqslant 0$ everywhere, then f is a constant function (and $\Delta f = 0$).*

Let $X = \text{grad } f$. Then

$$\int_M \Delta f dV = \int_M \text{div } X dV = 0$$

by Green's Theorem, since M has no boundary. Since $\Delta f \geqslant 0$ everywhere, it follows that $\Delta f = 0$ everywhere on M. Now

$$\int_M \Delta(f^2) dV = 2 \int f\Delta f dV + 2 \int (g^{ij} f_{,i} f_{,j}) dV$$

and

$$\int_M \Delta(f^2)dV = 0 \quad \text{by Green's Theorem.}$$

Since we have proved $\Delta f = 0$, we have

$$\int (g^{ij}f_{,i}f_{,j})dV = 0 \ .$$

Since the integrand is non-negative, we must have

$$g^{ij}f_{,i}f_{,j} = 0$$

from which $df = 0$ and hence f is a constant function. We have therefore proved that *on a closed manifold M, the only functions which satisfy $\Delta f = 0$ are constant functions.*

We now prove

Theorem 46. *There does not exist a closed minimal submanifold in a euclidean space.*

We now give an alternative proof of this theorem because the new proof uses ideas which will be useful later in more general conditions.

Let M be an n-dimensional Riemannian manifold immersed in euclidean space E^{n+p}. Since the computations are local, we may identify a point $x \in M$ with the position vector x of the corresponding point in E^{n+p}. Denote by a any constant vector in E^{n+p}; (we shall usually take a to be the ith vector of a standard orthonormal basis $e_1, \ldots, e_{n+p}$ of E^{n+p}, so that the function $f(x) = \langle x, a \rangle$ on M is just the ith-coordinate function of the immersion). We compute Δf when $f(x) = \langle x, a \rangle$.

Let Y be any vector field defined on M. Then

$$Yf = Y\{\langle x, a \rangle\} = \langle Y, a \rangle \ ,$$

where the Y of the right-hand side is the R^{n+p}-valued function which represents the vector field Y on M. Let X be a vector field on M. Then

$$XYf = \langle \nabla'_X Y, a \rangle$$

where ∇'_X is the covariant derivative with respect to the euclidean connexion of E^{n+p}. Now the second fundamental form α of M is related to the connexion ∇'_X and the induced connexion on M by the Gauss formula

$$\nabla'_X Y = \nabla_X Y + \alpha(X, Y) \ .$$

Since

$$(\nabla_X Y)f = \langle \nabla_X Y, a \rangle \ ,$$

we get

$$XYf-(\nabla_X Y)f=<\alpha(X, Y), a> .$$

Now the left-hand side of this equation is equal to

$$(\nabla^2 f)(Y;X) .$$

Let $\{X_1, \ldots, X_n\}$ be an orthonormal basis of the tangent space to M at x. Then

$$\Delta f=\sum_{i,j=1}^{n} g^{ij}(\nabla^2 f)(X_i;X_j)$$

$$=\sum_{i=1}^{n}(\nabla^2 f)(X_i;X_i) ,$$

so we get

$$\Delta f=\sum_{i=1}^{n}<\alpha(X_i, X_i), a> .$$

Let $\xi_1, \ldots, \xi_p$ be an orthonormal basis of the normal space at x. Then in terms of our previous notation we have

$$\alpha(X, Y)=\sum_{k=1}^{p} h^k(X, Y)\xi_p .$$

Thus

$$\frac{1}{n}\sum_{i=1}^{n}\alpha(X_i, X_i)$$

is just the mean curvature vector H at x. Thus we have proved the result

$$\Delta f=n<H, a> . \tag{5.25}$$

Corollary 47. *There exist no compact minimal submanifolds in a euclidean space.*

This follows because if we take $a=e_A$, $A=1, \ldots, n+p$, we see that when $H=0$, the Ath coordinate function of the immersion f must be a harmonic function and therefore a constant because M is compact. Thus M reduces to a single point.

We note that the corollary also follows immediately from theorem 43 for the special case $n=2$, $p=1$.

5.6 MINIMAL SUBMANIFOLDS OF SPHERES

We have already proved that there do not exist compact minimal submanifolds of euclidean space E^{n+p}. However, we shall now prove that such submanifolds do exist when we replace E^{n+p} by S^{n+p}. We shall establish this in two different ways—first by using the notation favoured by Kobayashi and Nomizu (1969), p. 341, and secondly by the method of exterior forms as used by Chern.

Let M^n be an n-dimensional submanifold of $S^{n+p}(r)$, that is, the $(n+p)$-dimensional sphere of radius r in euclidean space E^{n+p+1}. The radial vector $\xi = x/r$ is a unit vector at the point x on $S^{n+p}(r)$; let $\xi_1, \xi_2, \ldots, \xi_p$ together with ξ be an orthonormal frame which spans the normal space at x with respect to E^{n+p+1}. We denote by ∇', ∇^0 and ∇ the Riemannian connexions of E^{n+p+1}, S^{n+p} and M^n respectively. Let X, Y be vector fields tangent to M. Then we have

$$\nabla'_X Y = \nabla^0_X Y + h(X, Y)\xi \ , \tag{5.26}$$

where h is the second fundamental form of S^{n+p} relative to E^{n+p+1}. Moreover, we have

$$\nabla^0_X Y = \nabla_X Y + \sum_{\alpha=1}^{p} h^\alpha(X, Y)\xi_\alpha \tag{5.27}$$

where $h^1, \ldots, h^p$ are the second fundamental forms of M^n relative to S^{n+p}. Thus we have

$$\nabla'_X Y = \nabla_X Y + h(X, Y)\xi + \sum_{\alpha=1}^{p} h^\alpha(X, Y)\xi_\alpha \ . \tag{5.28}$$

Let $f(x) = <x, a>$ where a is a fixed vector in E^{n+p+1}. Using a similar argument to that of the previous section we see that

$$\Delta f = < \sum_{i=1}^{n} h(x_i, x_i)\xi + \sum_{i=1}^{n} \sum_{\alpha=1}^{p} h^\alpha(x_i, x_i)\xi^\alpha, a> \tag{5.29}$$

where $(x_1, \ldots, x_n)$ is an orthonormal basis of the tangent space to M^n at x. Clearly

$$\sum_{i=1}^{p} h(x_i, x_i) = -n/r^2 \tag{5.30}$$

and

$$\sum_{i=1}^{n} \sum_{\alpha=1}^{p} h^\alpha(x_i, x_i)\xi_\alpha$$

is equal to nH, where H is the mean curvature vector of M^n in S^{n+p}. In particular, the coordinate functions x^A considered as functions on M satisfy

$$\Delta x^A = -\left(\frac{n}{r^2}\right) x^A + nH^A \ . \tag{5.31}$$

Thus it follows that M^n is a minimal submanifold of S^{n+p} if and only if

$$\Delta x^A = -\left(\frac{n}{r^2}\right) x^A \ . \tag{5.32}$$

Example. Let (x, y, z) be rectangular cartesian coordinates in E^3 and let (u^i), $i=1, \ldots, 5$ be standard coordinates in E^5. Consider the mapping

$$u^1 = \frac{yz}{\sqrt{3}} \ , \quad u^2 = \frac{zx}{\sqrt{3}} \ , \quad u^3 = \frac{xy}{\sqrt{3}}$$

$$u^4 = \frac{1}{2\sqrt{3}}(x^2 - y^2) \ , \quad u^5 = \frac{1}{6}(x^2 + y^2 - 2z^2) \ .$$

Then a straightforward calculation gives

$$\sum_{i=1}^{5} (u^i)^2 = \left(\frac{x^2 + y^2 + z^2}{3}\right)^2$$

Thus the sphere S^2 with equation $x^2 + y^2 + z^2 = 3$ is mapped into the unit sphere S^4 given by

$$\sum_{i=1}^{5} (u^i)^2 = 1 \ .$$

Moreover the two points (x, y, z), $(-x, -y, -z)$ of S^2 map into the same point of S^4. Thus we can interpret the mapping as giving an imbedding of the real projective plane $P^2(\mathbb{R})$ into S^4. The image is called the **Veronese surface.** We show that this is a minimal surface of S^4.

To verify the above example, it is sufficient to prove that

$$\Delta u^i = -2u^i \ , \quad \text{for } i = 1, \ldots, 5 \ .$$

We illustrate by taking $i=1$.

We write

$$x = \sqrt{3} \sin\theta \cos\phi \ , \quad y = \sqrt{3} \sin\theta \sin\phi \ , \quad z = \sqrt{3} \cos\theta \ .$$

On the sphere $S^2(\sqrt{3})$ we have

$$ds^2 = 3(d\theta^2 + \sin^2\theta \, d\phi^2) \ .$$

Then $u^1 = \frac{\sqrt{3}}{2} \sin 2\theta \cos \phi$.

$$\Delta u^1 = \frac{1}{\sqrt{3} \sin \theta} \left[\frac{\partial}{\partial \theta} \left\{ \sqrt{3} \sin \theta \frac{1}{3} \frac{\partial}{\partial \theta} \left(\frac{\sqrt{3}}{2} \sin 2\theta \sin \phi \right) \right\} \right]$$

$$+ \frac{1}{\sqrt{3} \sin \theta} \left[\frac{\partial}{\partial \phi} \left\{ \sqrt{3} \sin \theta \frac{1}{3 \sin \theta} \frac{\partial}{\partial \phi} \left(\frac{\sqrt{3}}{2} \sin 2\theta \sin \phi \right) \right\} \right] .$$

A straightforward simplification of the right-hand side gives $-\sqrt{3} \sin 2\theta \sin \phi$ which is precisely $-2u^1$, as required.

We leave it to the reader to check that $\Delta u^i = -2u^i$ for $i = 2, 3, 4$ and 5.

5.7 MINIMAL SUBMANIFOLDS VIA EXTERIOR FORMS

It is interesting to give an analogous treatment of minimal sub-manifolds on the sphere by means of exterior forms.

Let $x : M^n \to S^{n+p}$ be a submanifold of $S^{n+p}(1)$ which itself is a unit sphere in E^{n+p+1}. Let (e_A) be an orthonormal frame of tangent vectors to S^{n+p} at x. Then (x, e_A) is an orthonormal frame in E^{n+p+1} satisfying

$$<x, x> = 1 , \quad <x, e_A> = 0 , \quad <e_A, e_B> = \delta_{AB} , \tag{5.33}$$

where the scalar product $<, >$ is defined for vectors in E^{n+p+1}.

We have

$$dx = \sum_A \omega_A \otimes e_A , \tag{5.34}$$

Let

$$de_A = \sum \omega_{AB} \otimes e_B + f_A \otimes x . \tag{5.35}$$

Then we have

$$<dx, e_A> + <x, de_A> = 0 ,$$

from which

$$<x, de_A> = -\omega_A .$$

This gives $f_A = -\omega_A$ so we get

$$de_A = \sum \omega_{AB} \otimes e_B - \omega_A \otimes x . \tag{5.36}$$

Also we have

$$\omega_{AB}+\omega_{BA}=0 \ . \tag{5.37}$$

Exterior differentiation of (5.36) gives

$$0=\Sigma d\omega_{AB}\otimes e_B-\Sigma\omega_{AB}\otimes de_B-d\omega_A\otimes x+\omega_A\otimes dx \ ,$$

from which we deduce that

$$0=\sum_C d\omega_{AB}\otimes e_B-\Sigma\omega_{AC}\omega_{CB}\otimes e_B+\Sigma\,\omega_A\wedge\omega_B\otimes e_B \ ,$$

that is

$$d\omega_{AB}-\Sigma\omega_{AC}\wedge\omega_{CB}=-\omega_A\wedge\omega_B \ . \tag{5.38}$$

The expression on the left-hand side of this equation gives the curvature form of the Riemannian metric on S^{n+p}. The components of the Riemann-Christoffel curvature tensor are given by

$$R_{ABCD}=\delta_{AC}\delta_{BD}-\delta_{AD}\delta_{BC} \ .$$

We choose adapted frames so that e_i are tangent vectors to the submanifold M. We denote the restrictions of the forms ω_A, ω_{AB} to M by θ_A, θ_{AB} respectively. Then equations (5.34), (5.36) become

$$dx=\sum_A\theta_A\otimes e_A \ ,$$

$$de_A=\sum_B\theta_{AB}\otimes e_B-\theta_A\otimes x \ ,$$

with $\theta_\alpha=0$, $\alpha=n+1,\ldots,n+p$.

The θ_{ij} are connexion forms of the induced metric on M so that its curvature forms are

$$\begin{aligned}d\theta_{ij}-\sum_k\theta_{ik}\wedge\theta_{kj}&=-\sum_\alpha\theta_{i\alpha}\wedge\theta_{j\alpha}-\theta_i\wedge\theta_j\\&=\tfrac{1}{2}\sum_{k,l,\alpha}(h_{i\alpha k}h_{j\alpha l}-h_{i\alpha l}h_{j\alpha k})\theta_k\wedge\theta_l-\theta_i\wedge\theta_j \ .\end{aligned}$$

The Riemann-Christoffel tensor of M therefore has components given by

$$S_{ijkl}=\sum_\alpha(h_{i\alpha k}h_{j\alpha l}-h_{i\alpha l}h_{j\alpha k})+\delta_{ik}\delta_{jl}-\delta_{il}\delta_{jk} \ . \tag{5.39}$$

This is effectively the Gauss-equation relating the curvatures of M and S. The Ricci curvature of M is given by

$$S_{ik}=-\sum_{\alpha,j}h_{i\alpha j}h_{k\alpha j}+(n-1)\delta_{ik} \ , \tag{5.40}$$

and the scalar curvature by

$$S=-\sum_{\alpha} h_{i\alpha j}^2+n(n-1)\ . \tag{5.41}$$

We set

$$\sigma=\sum_{\alpha,\,i,j} h_{i\alpha j}^2 \tag{5.42}$$

so that σ is the square of the length of the second fundamental form.

Let a denote a fixed unit vector in E^{n+p+1}, and consider the height function $<a, x>$ as a real-valued function on M. We have

$$d<x, a>=\Sigma<a, e_i>\theta_i\ , \tag{5.43}$$

$$D<a, e_i>=\Sigma\, h_{i\alpha j}(a, e_\alpha)\theta_j-<x, a>\theta_i\ . \tag{5.44}$$

We remember that the mean curvature vector of M^n in S^{n+p} is given by

$$H=\frac{1}{n}\sum_{i,\,\alpha} h_{i\alpha i}e_\alpha\ . \tag{5.45}$$

Then we have $\Delta<a, x>=\Sigma<a, e_\alpha>h_{i\alpha i}=n<a, H>-n<x, a>$. If we denote by a the unit vector parallel to e_A in E^{n+p+1} we get

$$\Delta x^i=n<a, H>-nx^i\ . \tag{5.46}$$

When $H=0$, this reduces to

$$\Delta x^i=-nx^i\ , \tag{5.47}$$

which agrees with the condition obtained by the alternative method used at the beginning of this section.

We shall make use of these results later on.

Example. Let E^{n+2} be considered as the direct sum

$$E^{n+2}=E_1^{r+1}\otimes E_2^{s+1}\ ,\quad r+s=n\ .$$

Thus a vector $\xi\in E^{n+2}$ may be uniquely written as

$$\xi=\xi_1+\xi_2\ ,$$

where $\quad \xi_1\in E_1^{r+1}\ ,\quad \xi_2\in E_2^{s+1}\ .$

We define the scalar product in E^{n+2} by

$$<\xi_1+\xi_2,\ \eta_1+\eta_2>=<\xi_1,\ \eta_1>+<\xi_2,\ \eta_2>\ . \tag{5.48}$$

Let ξ_λ be an arbitrary unit vector in E_λ, where $\lambda=1, 2$. Then the set of vectors $a_1\xi_1+a_2\xi_2$ with a_1, a_2 positive will describe a submanifold of dimension n on S^{n+1} provided that $a_1^2+a_2^2=1$. Consider the unit vector at $a_1\xi_1+a_2\xi_2$ given by

$$e_{n+1}=-a_2\xi_1+a_1\xi_2 \ .$$

Clearly e_{n+1} is a unit vector: moreover it is orthogonal to $a_1\xi_1+a_2\xi_2$. Also $<d\xi_1, -a_2\xi_1+a_1\xi_2>=0$ and $<d\xi_2, -a_2\xi_1+a_1\xi_2>=0$. Hence e_{n+1} is normal to M at $a_1\xi_1+a_2\xi_2$.

The second fundamental form of M is

$$-<dx, de_{n+1}>=a_1a_2(<d\xi_1, d\xi_1>-<d\xi_2, d\xi_2>) \ .$$

The induced metric on M is

$$a_1^2<d\xi_1, d\xi_1>+a_2^2<d\xi_2, d\xi_2> \ .$$

It follows that M is a minimal submanifold if and only if

$$\frac{r}{s}=\frac{a_1^2}{a_2^2} \ .$$

Thus we get a whole series of examples where a product of spheres $S^r\times S^s$ is embedded as a minimal submanifold of S^{r+s+1}. The special case $r=s=1$ gives the **Clifford torus** on S^3. We shall return to this special case later.

5.8 VARIATIONAL PROBLEMS OF HYPERSURFACES IN EUCLIDEAN SPACE

One method of considering the infimum of $\int_{M^2} H^2 dS$ is to apply standard techniques of the calculus of variations. We have already used this method in the section dealing with minimal submanifolds.

Let M^2 be an oriented surface of E^3, given by the vector-valued function $\mathbf{x}(u_1, u_2)$. The curves $u_1=$constant, $u_2=$constant are the parametric curves of the surface. The outer unit normal $\mathbf{N}$ at $\mathbf{x}$ is given by $\mathbf{x}_1\times\mathbf{x}_2/|\mathbf{x}_1\times\mathbf{x}_2|$ where $\mathbf{x}_i=\partial\mathbf{x}/\partial u_i$, $i=1, 2$. Let $g_{ij}=<\mathbf{x}_i, \mathbf{x}_j>$. Then the first fundamental form on M^2 is given by

$$I=\sum_{i,j} g_{ij}du_idu_j \ ;$$

the second fundamental form is given by

$$II=-<d\mathbf{N}, d\mathbf{x}>=\sum_{i,j} h_{ij}du_idu_j$$

where

$$h_{ij} = -\langle \mathbf{N}_i, \mathbf{x}_j \rangle = h_{ji} \ , \quad \mathbf{N}_i = \partial \mathbf{N}/\partial u_i \ .$$

As usual we denote by (g^{ij}) the inverse matrix of (g_{ij}). The mean curvature vector is given by

$$\mathbf{H} = \tfrac{1}{2} \sum_{i,j} g^{ij} h_{ij} \mathbf{N}$$

and the volume element by

$$dS = \sqrt{(\det (g_{ij}))}\, du_1 \wedge du_2 \ .$$

The equations of Gauss are of the form (cf. Willmore (1959), p. 112 (9.1))

$$\mathbf{x}_{ij} = \Sigma \Gamma^k_{ij} \mathbf{x}_k + h_{ij} \mathbf{N} \tag{5.49}$$

where $\mathbf{x}_{ij} = \partial^2 \mathbf{x}/\partial u_i \partial u_j$ and the Γ^k_{ij} are the Christoffel symbols given by

$$\Gamma^k_{ij} = \tfrac{1}{2} \sum_h g^{hk} (\partial g_{ih}/\partial u_j + \partial g_{jh}/\partial u_i - \partial g_{ij}/\partial u_h) \ .$$

The Weingarten equations (Willmore (1959), p. 113 (9.15)), give

$$\mathbf{N}_i = -\Sigma\, h^j_i \mathbf{x}_j$$

where we have written

$$h^j_i = \sum_k g^{jk} h_{ki} \ .$$

Consider the normal variation given by

$$\mathbf{x}(u_1, u_2, t) = \mathbf{x}(u_1, u_2) + t\phi(u_1, u_2) \mathbf{N}$$

where ϕ is a smooth real-valued function and t is a parameter in the open interval $(-\tfrac{1}{2}, \tfrac{1}{2})$.

We denote by δ the operator $\left.\dfrac{\partial}{\partial t}\right|_{t=0}$. We say that M is a *stable surface* if

$$\delta \int_{M^2} H^2 dS = 0 \ .$$

We have

$$\delta \mathbf{x} = \phi \mathbf{N} \ , \quad \delta \mathbf{x}_i = \phi_i \mathbf{N} + \phi \mathbf{N}_i \ .$$

$$\bar{g}_{ij} = \langle \mathbf{x}_i + t\phi_i \mathbf{N} + t\phi \mathbf{N}_i, \mathbf{x}_j + t\phi_j \mathbf{N} + t\phi \mathbf{N}_j \rangle$$

$$=g_{ij}+t\phi<\mathbf{x}_i,\mathbf{N}_j>+t^2\phi_i\phi_j+t\phi<\mathbf{N}_i,\mathbf{x}_j>+t^2\phi^2<\mathbf{N}_i,\mathbf{N}_j>\ .$$

Thus

$$\delta g_{ij}=-\phi\Sigma g_{ik}h_j^k-\phi\Sigma g_{jk}h_i^k=-2\phi h_{ij}\ .$$

Since

$$\Sigma g_{ij}.g^{jk}=\delta_i^k\ ,$$

we have

$$\Sigma\ \delta g_{ij}.g^{jk}+\Sigma\ g_{ij}\delta g^{jk}=0:$$

$$\therefore\ \Sigma\ g_{ij}\delta g^{jk}=2\Sigma g^{jk}\phi h_{ij}\ ,$$

and hence

$$\delta g^{ij}=2\phi\sum_k g^{jk}h_k^i\ .$$

Let

$$W=\sqrt{(\det(g_{ij}))},\text{ so that}$$

$$W^2=\det(g_{ij})\ .$$

Then

$$2W\frac{\partial W}{\partial t}=\sum_{i,j}\frac{\partial g_{ij}}{\partial t}\cdot W^2g^{ij}$$

$$\therefore\quad \delta W=\tfrac{1}{2}\delta g_{ij}.W.g^{ij}$$

$$=-\tfrac{1}{2}2\phi Wh_{ij}g^{ij}\ ,\quad\text{giving}$$

$$\delta W=-2\phi HW\ .$$

Moreover, since

$$<\mathbf{N},\mathbf{x}_i>=0\ ,\quad\text{we have}$$

$$<\delta\mathbf{N},\mathbf{x}_i>+<\mathbf{N},\delta\mathbf{x}_i>=0\ ,$$

and hence

$$<\delta\mathbf{N},\mathbf{x}_i>=-<\mathbf{N},(\phi_i\mathbf{N}+\phi\mathbf{N}_i)>$$

$$=-\phi_i\ .$$

Since

$$\mathbf{N}^2 = 1 \text{ we have } <\delta\mathbf{N}, \mathbf{N}> = 0 \ .$$

Write

$$\delta\mathbf{N} = \sum_j b^j \mathbf{x}_j \ .$$

Then

$$<\delta\mathbf{N}, \mathbf{x}_i> = \Sigma g_{ij} b^j = -\phi_i \ ,$$

giving

$$b^i = -\Sigma g^{ij}\phi_j \ , \quad \text{and hence}$$

$$\delta\mathbf{N} = -\Sigma g^{ij}\phi_j \mathbf{x}_i \ .$$

Using equation (5.49) we have

$$\delta\mathbf{N}\cdot\mathbf{x}_{ij} = -\sum_{p,q,k} <g^{pq}\phi_q\mathbf{x}_p, \Gamma_{ij}^k\mathbf{x}_k> \ , \tag{5.50}$$

$$d\mathbf{N}\cdot\mathbf{x}_{ij} = -\Sigma\phi_k\Gamma_{ij}^k \ . \tag{5.51}$$

We now wish to obtain the formula

$$<\mathbf{N}, \delta\mathbf{x}_{ij}> = \phi_{ij} - \phi\sum_k h_i^k h_{jk} \ . \tag{5.52}$$

We have

$$\bar{\mathbf{x}} = \mathbf{x} + t\phi\mathbf{N} \ , \quad \text{from which}$$

$$\bar{\mathbf{x}}_i = \mathbf{x}_i + t\{\phi_i\mathbf{N} + \phi\mathbf{N}_i\}$$

$$\bar{\mathbf{x}}_{ij} = \mathbf{x}_{ij} + t\{\phi_{ij}\mathbf{N} + \phi_j\mathbf{N}_i + \phi_i\mathbf{N}_j + \phi\mathbf{N}_{ij}\} \ .$$

Thus

$$\delta\mathbf{x}_{ij} = \phi_{ij}\mathbf{N} + \phi_j\mathbf{N}_i + \phi_i\mathbf{N}_j + \phi\mathbf{N}_{ij} \ .$$

The component of $\delta\mathbf{x}_{ij}$ along the normal is thus

$$\begin{aligned} &\phi_{ij} - \phi <\mathbf{N}_i, \mathbf{N}_j> \\ &\quad = \phi_{ij} - \phi <\Sigma h_i^p\mathbf{x}_p, h_j^q\mathbf{x}_q> \\ &\quad = \phi_{ij} - \phi h_{ik}h_j^k \ , \end{aligned}$$

which gives equation (5.52)

From (5.50) and (5.52) we obtain

$$\delta h_{ij} = \phi_{ij} - \sum_k \phi_k \Gamma^k_{ij} - \phi \sum_k h^k_i h_{kj} \ . \tag{5.53}$$

We may replace the terms $\phi_{ij} - \Sigma \phi_k \Gamma^k_{ij}$ by the covariant derivative $\nabla_i \nabla_j \phi$. Then

$$\delta h_{ij} = \nabla_i \nabla_j \phi - \phi \sum_k h^k_i h_{kj} \ . \tag{5.54}$$

We are now ready to apply the operation δ to our integral.

We have

$$\begin{aligned} \delta H &= \delta(g^{ij} h_{ij}) = \delta g^{ij} h_{ij} + g^{ij} \delta h_{ij} \\ &= 2\phi g^{jk} h^i_k h_{ij} + g^{ij} [\nabla_i \nabla_j \phi - \phi h^k_i h_{kj}] \\ &= \Delta \phi + \phi h^i_k h^k_i \ . \end{aligned} \tag{5.55}$$

Now the matrix $h = (h^i_k)$ has as eigenvalues the principal curvatures κ_1, κ_2. Hence

$$h^i_k h^k_i = \text{trace } \mathrm{h}^2 = \kappa_1^2 + \kappa_2^2 = (\kappa_1 + \kappa_2)^2 - 2\kappa_1 \kappa_2 \ .$$

Remembering that $\kappa_1 \kappa_2 = K$, the Gaussian curvature, we get

$$\delta H = \Delta \phi + \phi(4H^2 - 2K) \ .$$

Thus

$$\begin{aligned} \delta \int_{M^2} H^2 dS &= \int_{M^2} 2H \delta H dS + \int_{M^2} H^2 \delta dS \\ &= \int_{M^2} 2H \{\Delta \phi + \phi(4H^2 - 2K)\} dS - 2 \int_{M^2} H^2 \phi H dS \ . \end{aligned}$$

Since M^2 is a closed surface, it follows from Green's theorem that

$$\int_{M^2} H \nabla \phi dS = \int \phi \Delta H dS \ .$$

Thus we get

$$\delta \int_{M^2} H^2 dS = 2 \int_{M^2} \phi \{\Delta H + 2H(H^2 - K)\} dS \ .$$

The condition that the integral is stationary for all smooth functions ϕ is clearly

$$\Delta H + 2H(H^2 - K) = 0 \ . \tag{5.56}$$

This condition is clearly necessary if $\int H^2 dS$ attains a stationary value. However, to show that this corresponds to the infimum value requires an investigation of the second variation and this leads us outside the scope of the present book.

The history of equation (5.56) is interesting. I lectured on this topic at Oberwolfach, Germany, in the mid-1960's and thought it was new. However, I was informed by K. Voss that this was known to him in the mid-1950's, although he did not publish it. Both Voss and I were amused to find out subsequently that this equation was discovered by Thomsen (1923) and it is referred to explicitly by Blaschke in volume III of his text on differential geometry in 1929.

The formula was generalized by B-Y. Chen, "On a variational problem on hypersurfaces", *J. Lond. Math. Soc.* (2), **6** (1973), 321–325. Chen considered an m-dimensional oriented closed hypersurface in a euclidean space E^{m+1}. Let α be the mean curvature. There he proved that

$$\int_{M^m} |\alpha|^m dV \geqslant c_m \ , \tag{5.57}$$

where c_m denotes the surface area of the unit m-sphere in E^{m+1}. He also proved that the hypersurface is stable with respect to the integral

$$\int_{M^m} |\alpha|^m dV \tag{5.58}$$

if

$$\Delta\alpha^{m-1} + m(m-1)\alpha^{m+1} - \alpha^{m-1}R = 0 \ , \tag{5.59}$$

where R is the scalar curvature given by

$$R = m^2\alpha^2 - h_j^i h_i^j \ .$$

Equation (5.59) reduces to our equation (5.56) when $m=2$. In that case $R=2K$.

The method of Chen is almost identical to that given here for the case $m=2$.

5.9 VARIATIONAL PROBLEMS OF HYPERSURFACES IN RIEMANNIAN MANIFOLDS

One obvious way of extending this theory is to consider the variation of an integral involving the mean curvature of an n-dimensional manifold immersed as a hypersurface of a *general Riemannian manifold*. Some results in this direction were obtained by Willmore and Jhaveri (1972). Another generalization is to consider the immersion of 2-dimensional surfaces in a general Riemannian manifold. A good survey paper on this topic is given by Joel L. Weiner, "On a problem of Chen, Willmore, et al.", *Indiana University Math. Journal*, **27**, 1978, 19–35, who pays particular attention to solutions of the equation

$$\Delta H + 2H(H^2 - K) = 0 \ .$$

Clearly this equation is satisfied by spheres, when $\Delta H = 0$ and $H^2 - K = 0$ because every point is an umbilic. The reader should verify that it is also satisfied by the standard torus imbedded in E^3 by

$$x = (a + b\cos u)\cos v\ , \quad y = (a + b\cos u)\sin v\ , \quad z = a\sin u$$

if and only if $a/b = \sqrt{2}$.

Moreover, as was first explicitly proved by B.-Y. Chen (1974), the equation itself is invariant under conformal transformations of E^3.

Let M^2 be a closed orientable surface and let $f: M^2 \to S^3(1)$ be a smooth immersion of M^2 into $S^3(1)$, considered as a unit hypersphere of E^4. Let $G(m)$ be the value of the determinant of the second fundamental form of M^2 with respect to $S^3(1)$; let $K(m)$ denote the Gaussian curvature of M^2 at m, and let $\bar{K}(m)$ be the sectional curvature in S^3 of the 2-plane determined by $f_*(TM)_m$. Then the Gauss-equation gives

$$K(m) = G(m) + (\bar{K}m)\ . \tag{5.60}$$

The integral considered by Weiner in the general case $M^2 \to \bar{M}$ is

$$\tau(f) = \int_{M^2} (H^2 + \bar{K})dS \tag{5.61}$$

where $\bar{K}$ is defined as the sectional curvature on $\bar{M}$ of the 2-plane determined by $f_*(TM)_m$. This is also shown to be invariant under conformal changes of metric of $\bar{M}$. A key result of this paper is that *any closed minimal surface in $\bar{M}$ is a stationary point of τ*. In the case $\bar{M} = S^3$ the condition for τ to be stationary is

$$\Delta H + 2H(H^2 - G) = 0\ , \tag{5.62}$$

where G is the determinant of the second fundamental form relative to S^3. Naturally when $\bar{M} = E^3$ the same equation holds with $G = K$. We can map S^3 into E^3 by stereographic projection σ from the north pole onto the equatorial space, and since this is a conformal map we have $\tau(f) = \tau(\sigma \circ f)$. Thus we have

Theorem 48. *Let $f: M^2 \to S^3$ be an immersion of the closed orientable surface M^2 into the unit sphere S^3. Let σ be stereographic projection $S^3 \to E^3$. Then f satisfies equation (5.62) in S^3 if and only if $\sigma \circ f$ satisfies (5.56) in E^3.*

We now make use of a remarkable theorem due to H. B. Lawson (1968) proved incidentally in his Ph.D. Thesis:

Lawson's Theorem 49. *There exist closed orientable embedded minimal surfaces of arbitrary genus in S^3.*

This theorem has the immediate

Corollary 50. *There exist embedded surfaces of E^3 of arbitrary genus that satisfy*

$$\Delta H + 2H(H^2 - K) = 0 \ .$$

We also note that if *M^2 is immersed as a minimal surface of S^3, the value of $\tau(f)$ is equal to the area of $f(M^2)$*. This follows immediately because

$$\tau(\sigma \bigcirc f) = \tau(f) = \int_{M^2} (H^2 + 1) dS = \int_{M^2} dS$$

where we have used the fact that $\bar{K} = 1$.

The problem remains of finding a generalization of the integral

$$I = \int H^2 dS$$

for an n-dimensional closed manifold M immersed in a Riemannian manifold M' of dimension $n+p$, so that the integral is invariant under a conformal change of metric of M', but nevertheless the Euler equation corresponding to $\delta I = 0$ has a manageable form.

We have seen that the choice of integrand H^n used by B.-Y. Chen for hypersurfaces does lead to a manageable Euler equation but the integrand is not a conformal invariant. Let us restrict ourselves to the case of an n-dimensional manifold M immersed in a Riemannian manifold M' of dimension $n+1$. Consider the equation for the principal curvatures

$$\kappa^m - \sigma_1 \kappa^{m-1} + \sigma_2 \kappa^{m-2} - \ldots \pm \sigma_m = 0 \ , \tag{5.63}$$

where σ_i denotes the elementary symmetric polynomial of degree i of the roots. Consider the conformal change of metric of M' given by $M' : g \to \bar{g} = \lambda^2 g$.

Exercise

Show, by a straightforward calculation, that if l is the second fundamental form with respect to the normal direction N, then

$$\bar{l} = \lambda(l + \nabla_N \log \lambda \cdot g) \ . \tag{5.64}$$

It follows that

$$\bar{\kappa}_i - \bar{\kappa}_j = (\kappa_i - \kappa_j)/\lambda \ , \qquad i < j$$

with

$$d\bar{V} = \lambda^m dV \ .$$

We define the normalized coefficients of (5.63) by

$$H_i = \sigma_i / \binom{n}{i} \; .$$

Then we leave the reader to check:

Exercise

$$I = \int_M (H_1^2 - H_2)^{n/2} dV \tag{5.65}$$

is conformally invariant.

Computation of the variational equation corresponding to (5.65) gives the appropriate Euler equation. This was done independently by myself and by Karcher and Voss jointly during the summer of 1972. However, the form of the Euler equation obtained by Karcher and Voss seems to be more significant than mine, so I give below the precise form obtained by them in a lecture at Oberwolfach, September 1972.

Let

$$\psi = (H_1^2 - H_2)^{(n/2)-1} \; . \tag{5.66}$$

Let

$$d^{ij} = (n-1)^{-1}(l^{ij} - H_1 g^{ij}) \; . \tag{5.67}$$

Let

$$R = \text{curvature tensor of } M' \; , \quad \text{Ric} = \text{Ricci tensor of } M' \; .$$

Let X_i, $i = 1, 2, \ldots, n$ be an orthonormal frame of vector fields over M. Then the Euler equation corresponding to (5.65) can be written

$$0 = \psi[\Delta H_1 + n\{(n-1)H_1(H_1^2 - H_2) + \frac{n-2}{2}(H_3 - H_1 H_2)\} + d^{ij}R(N, X_i, N, X_j) - \text{Ric}\,(N, X^i)_{;i}] + 2(H_1^i - \text{Ric}\,(N, X^i))\psi_i + d^{ij}\psi_{ij} \; . \tag{5.68}$$

When written in this form, the awkward curvature terms disappear when M' has constant curvature. However, the real challenge seems to obtain results under less restrictive curvature assumptions about the metric on M'—for example, when M' has an Einstein metric. This remains one of many open problems.

5.10 KNOTTED TORI

It was proved by B.-Y. Chen that a knotted torus T in E^3 must satisfy the condition

$$\int_T H^2 dS \geqslant 8\pi \ . \tag{5.69}$$

We now obtain an improvement of this result by replacing the sign on the right-hand side by strict inequality. It has been proved by Kuiper and Meeks (1982) that a knotted torus T in E^3 satisfies

$$\int_T |K| dS > 16\pi \ . \tag{5.70}$$

Let us denote the regions of the torus T where K is positive (negative) by $K>$ $(K<)$ respectively. Then the above result gives

$$\int_{K>} KdS - \int_{K<} KdS > 16\pi \ . \tag{5.71}$$

By the Gauss-Bonnet theorem we have

$$\int_{K>} KdS + \int_{K<} KdS = 2\pi\chi(T) = 0 \ .$$

Thus we have

$$\int_{K>} KdS > 8\pi \ . \tag{5.72}$$

Now

$$\int_T H^2 dS \geqslant \int_{K>} H^2 dS \geqslant \int_{K>} KdS > 8\pi \ .$$

Thus we have proved that a knotted torus necessarily satisfies the condition

$$\int_T H^2 dS > 8\pi \ . \tag{5.73}$$

In recent unpublished woik C. Kearton has generalized the inequality involving $|K|$ to give

$$\int_T |K| dS \geqslant 16\pi n$$

where n is the bridge number. A similar argument to that given previously now yields that a knotted torus of bridge number n satisfies the inequality

$$\int_T H^2 dS \geqslant 8\pi n \ . \tag{5.74}$$

Chapter 6

Conformal Volume

6.1 INTRODUCTION

This chapter is based on a very recent work by Peter Li and Shing-Tung Yau (1982). I thank the authors for their generosity in allowing me to use this material before it is published. The importance of this approach is that it brings together the spectral theory of the Laplacian operator with the theory of conformal invariants, and in particular it has applications to the problem of computing the infimum of the integral of the square of the mean curvature of surfaces immersed in Riemannian manifolds.

Let (M, γ) be an m-dimensional compact Riemannian manifold. Let $S^n = S^n(1)$ denote the unit sphere regarded as a submanifold of euclidean space E^{n+1}. Let ϕ be a conformal map of M into the sphere S^n. Such a map may not exist for an arbitrarily given (M, γ), but we shall restrict attention only to those for which such a map exists. Of course, if we regard M as a compact differentiable manifold and use ϕ to induce a Riemannian metric from S^n to M, the existence of a ϕ is guaranteed.

Let us denote by ds_0^2 the standard metric on S^n, and by ds^2 the given metric on M. Since ϕ is a conformal map, we have

$$\phi^* ds_0^2 = \alpha(x) ds^2, \tag{6.1}$$

where $\alpha(x)$ is a positive real-valued function defined on M.

Consider now the set of diffeomorphisms of S^n which induce conformal changes of metric of S^n. It is clear that this set forms a group, which we denote by G. Moreover if we compose the conformal diffeomorphism $g \in G$ with ϕ, we get a conformal map $g \circ \phi$ of M to S^n. Each such map will pull back to M a metric derived from ds_0^2 on S^n. Let dV_g denote the volume element of the pull-back tensor $\phi^* \circ g^* ds_0^2$, and denote by

$$V_c(n, \phi, g) = \int_M dV_g \,, \tag{6.2}$$

the volume of M as measured by this pulled back metric. We take the supremum of this set of numbers corresponding to all values of g, and call this the **n-conformal volume of** ϕ, denoted by $V_c(n, \phi)$. More precisely we have

$$V_c(n, \phi) = \sup_{g \in G} \int_M dV_g \ . \tag{6.3}$$

We now take the infimum of $V_c(n, \phi)$ where ϕ runs over all non-degenerate conformal mappings of M into S^n, to obtain the ***n*-conformal volume of *M*.** More precisely

$$V_c(n, M) = \inf_{\phi} V_c(n, \phi) \ . \tag{6.4}$$

A natural question to ask at this stage is why not take the *infimum* of $V_c(n, \phi)$ for $g \in G$ and the *supremum* of the resulting set when ϕ runs over the space of non-degenerate conformal mappings. Clearly this would be undesirable because the last step may lead to a non-finite value.

The n-conformal volume of a Riemannian manifold is a new concept. In the particular case when M is a surface (orientable or non-orientable) we shall talk about the conformal area of M. This turns out to be a non-trivial invariant. We shall see, for example, that the conformal area of the real projective plane $P^2(\mathbb{R})$ is 6π while that for the square torus (Clifford torus) is $2\pi^2$. It appears that the computation of the conformal area for general surfaces will play in the future a significant role in studying the geometry of compact surfaces. In particular, as perhaps suggested by the number $2\pi^2$ associated with the square torus, there will be a close relation between conformal area of a surface and the integral of the square of the mean curvature.

We now show that $V_c(n, M)$ is non-increasing with n. More precisely, we have

Theorem 51. $V_c(n, M) \geq V_c(n+1, M)$.

If we regard S^n as an equatorial sphere of S^{n+1}, a conformal map $\phi : M \to S^n$ can be regarded as a conformal map into S^{n+1}. Since in defining $V_c(n+1, M)$ we take the infimum over *all* conformal maps and not merely those arising from S^n, we obtain the required inequality.

We can now define the **conformal volume** of M by the formula

$$V_c(M) = \lim_{n \to \infty} V_c(n, M) \ .$$

An important early remark is that if the manifold M admits a conformal map of degree d onto another manifold N, then

$$V_c(n, M) \leq |d| \, V_c(n, N) \ . \tag{6.5}$$

We remind the reader that our assumption about the degree implies that, in general, d points of M will map to the same point of N. Moreover, since $V_c(n, M)$ is defined by taking the infimum over *all* conformal maps from M to S^n, we clearly must have the inequality sign in (6.5). From (6.5) it follows that if we can compute the value of $V_c(n, N)$ and if we know that there exists a conformal map of degree d of another manifold M onto N, then we have an upper bound for $V_c(n, M)$. In applications we shall usually take N as S^2 or $P^2(\mathbb{R})$. We first prove the following

Theorem 52. *Let M be an m-dimensional Riemannian manifold. Then*

$$V_c(n, M) \geqslant V_c(n, S^m) = V(S^m) \ . \tag{6.6}$$

To prove this, let θ be a point on S^n. This will determine a unit vector field of parallel vectors in E^{n+1}, which we restrict to S^n. At each point p on S^n we project the vector of this field at p onto the tangent plane to S^n at p. In this way we get a field of tangent vectors which vanish both at θ and $-\theta$. This vector field determines a one parameter conformal subgroup of G, which we denote by $g_\theta(t)$, and this fixes the points θ and $-\theta$ for all values of t. Moreover, as t increases the effect of $g_\theta(t)$ is to move each point of S^n, except the point $-\theta$, towards θ. More precisely we have

$$\lim_{t\to\infty} g_\theta(t)(z) = \theta \tag{6.7}$$

for all $\quad z \in S^n \setminus \{-\theta\}$.

Let $x \in M$ and let $\phi : M \to S^n$ be a conformal map whose differential at x is of rank m. Then the volume of $g_{-\phi(x)}(t) \circ \phi(M)$ will tend to some non-trivial integral multiple of the volume of S^m as $t \to \infty$, the sphere S^m being the intersection of S^n with the m-dimensional subspace determined by the tangent space at $\phi(x)$. This shows that of all manifolds of dimension m, the sphere S^m gives rise to the smallest value of $V_c(n, M)$, and the theorem is therefore proved.

We now make use of a consequence of the famous Riemann-Roch theorem for compact surfaces without a boundary. In particular, it follows that if Σ is an orientable compact surface, without a boundary, of genus g, then there exists a conformal map onto S^2 with degree $\leqslant (g+1)$. Then

$$V_c(n, \Sigma) \leqslant (g+1)\, V_c(n, S^2) = (g+1)\, V(S^2) = 4(g+1)\pi \ . \tag{6.8}$$

Similarly if Σ is non-orientable, we can consider the orientable double cover $\tilde{\Sigma}$. It is known that we can choose a conformal map from $\tilde{\Sigma}$ onto S^2 which commutes with the deck transformation of $\tilde{\Sigma}$ and the antipodal map of S^2. This gives rise to a conformal map onto $P^2(\mathbb{R})$. We then get

$$V_c(n, \Sigma) \leqslant 2(g+1)\, V_c(n, P^2(\mathbb{R})) = 2(g+1)\, V(P^2(\mathbb{R})) \ . \tag{6.9}$$

In section 6.3 we shall prove that if $n \geqslant 4$, then

$$V_c(n, P^2(\mathbb{R})) = 6\pi \ . \tag{6.10}$$

This with the previous inequality gives

$$V_c(n, \Sigma) \leqslant 12(g+1)\pi \tag{6.11}$$

provided that $n \geqslant 4$.

6.1.1 An example

At this stage it might be helpful to calculate $V_c(4, \phi, g)$ from equation (6.2) when g is the identity and where ϕ is given by

$$\left.\begin{aligned} &u^1 = \frac{1}{\sqrt{3}} yz \;, \quad u^2 = \frac{1}{\sqrt{3}} zx \;, \quad u^3 = \frac{1}{\sqrt{3}} xy \;, \\ &u^4 = \frac{1}{2\sqrt{3}}(x^2 - y^2) \;, \quad u^5 = \frac{1}{6}(x^2 + y^2 - 2z^2) \;. \end{aligned}\right\} \tag{6.12}$$

Here $(u^1, u^2, u^3, u^4, u^5)$ is the standard coordinate system in E^5, (x, y, z) is the standard coordinate system in E^3 and we have $x^2+y^2+z^2=3$. A straightforward calculation gives

$$\begin{aligned} \sum_{i=1}^{5} (u^i)^2 &= \frac{1}{3}\Big[y^2z^2 + z^2x^2 + x^2y^2 + \frac{1}{4}(x^4 + y^4 - 2x^2y^2) + \\ &\qquad \frac{1}{12}(x^4 + y^4 + 4z^4 - 4x^2z^2 - 4y^2z^2 + 2x^2y^2) \Big] \\ &= \frac{1}{36}\Big[12y^2z^2 + 12z^2x^2 + 12x^2y^2 + 3x^4 + 3y^4 - 6x^2y^2 \\ &\qquad + x^4 + y^4 + 4z^4 - 4x^2z^2 - 4y^2z^2 + 2x^2y^2 \Big] \\ &= \frac{1}{36}\Big[4x^4 + 4y^4 + 4z^4 + 8y^2z^2 + 8z^2x^2 + 8x^2y^2 \Big] \\ &= \frac{1}{9}\Big[x^2 + y^2 + z^2 \Big]^2 = 1 \;. \end{aligned}$$

Hence we see that (6.12) gives a mapping of $S^2(\sqrt{3})$ into $S^4(1)$. Since the points (x, y, z) and $(-x, -y, -z)$ are mapped into the same point of $S^4(1)$, this defines an imbedding of $P^2(\mathbb{R})$ into $S^4(1)$. In fact, this is the Veronese surface that we have met before in Chapter 5.

We now compute the area of this surface.
From (6.12) we have

$$du^1 = \frac{1}{\sqrt{3}}(ydz + zdy) \;, \quad du^2 = \frac{1}{\sqrt{3}}(zdx + xdz) \;, \quad du^3 = \frac{1}{\sqrt{3}}(xdy + ydx) \;,$$

$$du^4 = \frac{1}{\sqrt{3}}(xdx - ydy) \ , \quad du^5 = \frac{1}{3}(xdx + ydy - 2zdz) \ .$$

A straight-forward calculation now gives

$$\sum_{i=1}^{5} (du^i)^2 = \frac{1}{3}\left[\begin{array}{l} y^2dz^2 + z^2dy^2 + 2yzdydz + z^2dx^2 + x^2dz^2 + 2xzdxdz + \\ x^2dy^2 + y^2dx^2 + 2xydxdy + x^2dx^2 + y^2dy^2 - 2xydxdy + \\ \frac{1}{3}\{x^2dx^2 + y^2dy^2 + 4z^2dz^2 + 2xydxdy - 4xzdxdz - \\ 4yzdydz\} \end{array}\right]$$

$$= \{(1 + \frac{x^2}{3})dx^2 + (1 + \frac{y^2}{3})dy^2 + (1 + \frac{z^2}{3})dz^2 +$$

$$\frac{2}{3}(dxdy + dydz + dzdx)\}$$

$$= \{dx^2 + dy^2 + dz^2 + \frac{1}{3}(xdx + ydy + zdz)^2\}$$

$$= \{dx^2 + dy^2 + dz^2\} \ ,$$

since on the sphere $S^2(\sqrt{3})$,

$$xdx + ydy + zdz = 0 \ .$$

If we write

$$x = \sqrt{3} \sin\theta \cos\phi \ ,$$

$$y = \sqrt{3} \sin\theta \sin\phi \ ,$$

$$z = \sqrt{3} \cos\theta \ ,$$

then

$$dx = \sqrt{3}(\cos\theta \cos\phi \, d\theta - \sin\theta \sin\phi \, d\phi) \ ,$$

$$dy = \sqrt{3}(\cos\theta \sin\phi \, d\theta + \sin\theta \cos\phi \, d\phi) \ ,$$

$$dz = -\sqrt{3} \sin\theta \, d\theta \ ,$$

and we get

$$\sum_{i=1}^{5} (du^i)^2 = 3(d\theta^2 + \sin^2\theta \, d\phi^2) \ .$$

The area of the image is therefore $\frac{1}{2}(3 \times 4\pi)$; the factor $\frac{1}{2}$ coming because we are dealing with $P^2(\mathbb{R})$ and not the sphere $S^2(\sqrt{3})$ which is the double covering.

We have thus obtained the value 6π for this embedding of $P^2(\mathbb{R})$ into S^4, which incidentally agrees with the value in (6.10).

6.2 THE SPECTRUM OF A RIEMANNIAN MANIFOLD

The best account of the spectrum of a Riemannian manifold is to be found in the lecture notes by Berger, Gauduchon and Mazet, *Le Spectre d'une Variété Riemannienne*, Lecture Notes No. 194, Springer-Verlag, (1971), and we refer the reader to this source for detailed proofs and other results. However, the following summary contains the matters particularly relevant to this Chapter, especially the properties of flat tori.

6.2.1 Definition of spectrum

Let (M, g) be a Riemannian manifold which is connected and compact. Let Δ be the *negative* of the classical Laplacian operator acting on the class of smooth real-valued functions $\mathscr{C}^\infty(M)$ defined on M. Specifically we have

$$\Delta f = -\sum_{i,j=1}^{n} \frac{1}{\sqrt{g}} \frac{\partial}{\partial x_i} \left(g_{ij} \sqrt{g} \frac{\partial f}{\partial x_j} \right) . \tag{6.13}$$

We denote by **Spec** (M, g) *the set of real numbers* λ *such that there exist a smooth function* f, $f \neq 0$, *satisfying* $\Delta f = \lambda f$.

Let us consider now the space of complex-valued smooth functions $\mathscr{C}^\infty_{\mathbb{C}}(M)$ on M. This can be given an inner product as follows:–

$$\langle f, h \rangle_{\mathbb{C}} = \int_M f\bar{h}\, dV .$$

Let $\Delta_{\mathbb{C}}$ be the extension of the operator Δ to $\mathscr{C}^\infty_{\mathbb{C}}(M)$. That is, if we write $f = u + iv$ where u and v are real-valued functions, then

$$\Delta_{\mathbb{C}} f = \Delta u + i\Delta v .$$

Clearly $\Delta_{\mathbb{C}}$ is a real operator which is self-adjoint and positive definite. So we can also regard **Spec** (M, g) as the set of numbers $\lambda \in \mathbb{C}$ such that there exists a function $f \in \mathscr{C}^\infty_{\mathbb{C}}(M)$, $f \neq 0$, satisfying

$$\Delta_{\mathbb{C}} f = \lambda f .$$

Each function $f \in \mathscr{C}^\infty(M)$ such that $\Delta f = \lambda f$, with $\lambda \in$ **Spec** (M, g) is called an **eigen-function associated with** λ. The set of eigen-functions associated with λ is a subspace of $\mathscr{C}^\infty(M)$ called the **eigen-subspace relative to** λ and is denoted by $\mathscr{P}_\lambda(M, g)$. The sum of all such subspaces, which we denote by $\mathscr{P}(M, g)$, that is

$$\mathscr{P}(M, g) = \sum_{\lambda \in Spec(M,g)} \mathscr{P}_\lambda(M, g) \tag{6.14}$$

is called the **eigenspace of (M, g)**

The decomposition given in (6.14) may be shown to be orthogonal.

Moreover, it may be proved that the spectrum of (M, g) has the following properties:–

(1) The spectum forms a discrete sequence $\{0=\lambda_0<\lambda_1<\lambda_2<\ldots\}$ tending to $+\infty$.
(2) For each $\lambda\in$**Spec** (M, g), $P_\lambda(M, g)$ is finite dimensional. The eigen-subspace associated with λ_i will be denoted by $P_i(M, g)$, and its dimension, denoted by m_i is called the **multiplicity** of λ_i.
(3) The **partition function** of (M, g), denoted by $Z(M, g)$ is defined by

$$Z(M, g)=\sum_{i=0}^{\infty} m_i e^{-\lambda_i t} \tag{6.15}$$

for $t>0$. Let μ be a positive real number and form the function $e^{\mu t}Z(t)$. Then it can be proved that λ_1 is the unique value of μ for which $e^{\mu t}[Z(t)-1]$ tends to a finite limit >0 as $t\to\infty$, and that this limit is in fact m_1. Moreover λ_i is the unique μ such that

$$e^{\mu t}[Z(t)-\sum_{j=0}^{i-1} m_j e^{-\lambda_j t}]$$

tends to a finite limit >0 as $t\to\infty$, and moreover, this limit is m_i. Thus the function $Z(t)$ determines the spectrum of (M, g).

6.2.2 The flat tori

Let $(\tilde{M}, g)$ be a Riemannian manifold and let G be a group of isometries of $(\tilde{M}, g)$, which operates in such a way that the projection $p\colon \tilde{M}\to M=M/G$ is a covering map. Then M inherits a canonical Riemannian metric. To see this, let $\tilde{m}$ be a fixed point in the fibre over $m\in M$. Since p is a local diffeomorphism, we may define $g_0(m)$ by

$$g_0(m)(X, Y)=g(\tilde{m})(p_*{}^{-1}X, p_*{}^{-1}Y)\ . \tag{6.16}$$

If we had chosen another point $\tilde{m}'$ in the fibre over m instead of $\tilde{m}$, then since $\tilde{m}'$ can be obtained from m by an application of an element of G, that is, by an isometry which preserves scalar product, we would clearly have obtained a result consistent with (6.16). Thus the induced metric g_0 is canonically determined.

We have already used this general result in the particular case when we regarded the sphere S^n as a double cover of the real projective space $P^n(\mathbb{R})$, to obtain a canonical metric on $P^n(\mathbb{R})$. We now use the same technique to obtain metrics on flat tori.

The **flat tori** are Riemannian manifolds obtained from n-dimensional euclidean space R^n with a flat (euclidean) metric by the action of free abelian groups, of maximal rank, of R^n. These manifolds are diffeomorphic to the torus $R^n/\mathbb{Z}^n$, where $\mathbb{Z}^n$ is the n-fold product of the abelian group of integers. These manifolds are compact.

We now show that *two flat tori are isometric if and only if there is an isometry of* R^n *which interchanges their defining subgroups.* Let Γ be the subgroup, regarded as a lattice, which defines the flat torus T. Similarly let Γ' define T'. Let p and p' be the corresponding canonical projections. Suppose that $f: T \to T'$ is an isometry. Let $m \in R^n$. The line segment joining the origin O of R^n to m will project under p into a curve in the torus T, passing through $p(O)$ and $p(m) = m_o$. The image of this curve by the map f is a curve in T' passing through $m_o' = f(m_o)$ which lifts uniquely to determine a unique point in R^n. We thus get a mapping $F: R^n \to R^n$ which covers the map f. F is a diffeomorphism. Moreover, since p and p' are local isometries and f is an isometry, then F is an isometry of R^n. The converse is trivially satisfied, and so our claim is justified.

In particular it follows that 2-dimensional flat tori are classified isometrically according to the lattices Γ in R^2. Let a be a non-zero element of Γ of minimal length, which after a suitable rotation we may think as located on the x-axis in R^2 from O to a. Clearly by a change of scale we may represent this element by the end point $(1, 0)$. The lattice is uniquely determined by the coordinates (x, y) of a second vector b of minimal length in $\Gamma/\{1, 0\}$. We can impose on (x, y) the restrictions $-\frac{1}{2} \leqslant x \leqslant \frac{1}{2}$, and $\sqrt{(1-x^2)} \leqslant y \leqslant 1$, since $\boldsymbol{b}$ must not have shorter length than $\boldsymbol{a}$.

6.2.3 Computation of λ_1 for the flat torus

Let Γ^* be the set of vectors $x \in R^n$ such that for every $y \in \Gamma$, the inner product $\langle x, y \rangle$ is an integer. Then it is not difficult to prove that Γ^* is a lattice, called the **dual lattice** of Γ. Moreover $(\Gamma^*)^* = \Gamma$. The proof consists of choosing a basis $\{e_i, \ldots, e_n\}$ for Γ. This is a basis for R^n and the dual basis $\{e_i^*, \ldots, e_n^*\}$ which satisfies $e_i^*(e_j) = \delta_{ij}$, serves as a basis for Γ^*. The fact that $(\Gamma^*)^* = \Gamma$ now follows, since the spaces concerned are finite-dimensional.

For each $x \in \Gamma^*$, we define a C^∞-function on R^n by

$$f_k(y) = exp(2\pi i x(y)) \,. \tag{6.17}$$

Since $x \in \Gamma^*$, this function determines a function f_x on the torus T determined by Γ. Let x^i, y^i be components of x and y referred to the bases (e_i^*) and (e_i). Then we have

$$f_x(y) = exp\left(2\pi i \sum_{j=1}^{n} x^j y^j\right) . \tag{6.18}$$

Differentiating with respect to y^j gives

$$\frac{\partial f_x(y)}{\partial y^j} = 2\pi i \, x^j f_x(y) \;,$$

and a second differentiation gives

$$\frac{\partial^2 f_x(y)}{(\partial y^j)^2} = -4\pi^2 (x^j)^2 f_x \ .$$

We interpret Δ as the negative of the classical Laplacian operator to get

$$\Delta f_x = 4\pi^2 \, \Sigma (x^j)^2 f_x = 4\pi^2 |x|^2 f_x \ . \tag{6.19}$$

We see immediately that $\lambda = 4\pi^2|x|^2$ is an eigenvalue of Δ corresponding to the eigenfunction f_x, with $x \in \Gamma^*$. To each value of λ is associated a subspace of eigenfunctions V_λ, which is generated by the f_x with $|x|^2 = \frac{\lambda}{4\pi^2}$. The dimension of V_λ is equal to the number of $x \in \Gamma^*$ such that $|x|^2 = \frac{\lambda}{4\pi^2}$. By an application of the Stone-Weierstrass theorem, it turns out that *all* the eigenvalues of Δ with corresponding eigenfunctions are precisely of this form. Thus, if we apply these calculations to the flat torus T^2 determined by the lattice (0, 1), (x, y) where $-\frac{1}{2} \leqslant x \leqslant \frac{1}{2}$, $\sqrt{1-x^2} \leqslant y \leqslant 1$, we find that the smallest non-zero eigenvalue λ_1 is given by

$$\lambda_1 = \frac{4\pi^2}{y^2} \tag{6.20}$$

Incidentally it is easy to prove that if two 1-dimensional Riemannian manifolds have the same spectra, then they are isomorphic. The same property holds for flat 2-dimensional tori; indeed, it can be proved that this also holds for any flat 2-dimensional manifolds. However, J. Milnor has constructed two tori of dimension 16 which are not isometric but nevertheless they have the same spectrum.

6.3 THE FIRST EIGENVALUE OF Δ AND THE CONFORMAL AREA FOR SURFACES

We now make use of the famous mini-max principle for the first non-zero eigenvalue λ_1 which in our case gives

$$\lambda_1 = inf \left[\int_M |\nabla f|^2 dV / \int f^2 dV \right] \tag{6.21}$$

where the infimum is taken over all continuous functions such that $\int_M f dV = 0$. (See, for example, Courant-Hilbert, *Methods of Mathematical Physics, Vol. I*, Interscience 1962, p. 399). The above formula seems to be well-known to engineers and physicists but, rather surprisingly, little known to mathematicians in general. It is related to the Rayleigh-Ritz method in engineering.

Let ϕ be a conformal map of a compact surface M into S^n, such that

$$V_c(n, \phi) \leqslant V_c(n, M) + \varepsilon \ . \tag{6.22}$$

Let X_i, $i=1, \ldots, n+1$ be coordinate functions of R^{n+1}. We now claim that there exists an element of $g \in G$, the conformal group of S^n, such that

$$\int_M (X_i \circ g \circ \phi) dV_g = 0 \ , \tag{6.23}$$

for all $i=1, \ldots, n+1$. The idea is to use these as trial functions in equation (6.21). We sketch the proof of the existence of such an element g, following closely the argument of Li and Yau.

We first note that the action of G on S^n can be extended to the unit ball B^{n+1} which is bounded by S^n in R^{n+1}. The isotropy subgroup at the origin of S^n is just $O(n+1)$. Now suppose that we have a point $A \in B^{n+1}$ with $A \neq O$. We obtain a vector field over R^{n+1} by parallel displacement of the unit vector $A/||A||$; we restrict this vector field to S^n and project onto the tangent space of each point of S^n. The conformal vector field so obtained extends to be a conformal vector field of B^{n+1}. This field generates a one-parameter family of conformal automorphisms $g(t)$ of B^{n+1} onto itself. Now in this group $g(t)$, there exists a unique conformal automorphism g_A which maps the origin to the point A. This construction gives rise to an embedding of B^{n+1} into G, which we denote by F.

We now define a map H in the reverse direction, that is mapping from $F(B^{n+1})$ to B^{n+1}. To each $g \in F(B^{n+1})$ we associate the point of B^{n+1} given by

$$H(g) = \frac{1}{V(M)} \int_M (X \circ g \circ \phi) dV_g \ . \tag{6.24}$$

Clearly the map HF maps B^{n+1} into itself, and when restricted to $\partial B^{n+1} = S^n$ it is just the identity map. As a consequence of the Brouwer fixed point theorem, it follows that HF must be surjective. Hence there must be some element g such that (6.23) is satisfied, and the existence of a suitable g is proved.

We note that for each i such that $1 \leqslant i \leqslant n+1$,

$$\int_M |\nabla X_i \circ g \circ \phi|^2 dV = \int_M (g \circ \phi)^* [|\nabla X_i|^2 dV_g] \ . \tag{6.25}$$

Therefore

$$\begin{aligned} \sum_{i=1}^{n+1} \int |\nabla X_i \circ g \circ \phi|^2 dV &= \int_M (g \circ \phi)^* [\Sigma |\nabla X_i|^2 dV_g] \\ &= 2 \int_M (g \circ \phi)^* dV_g \\ &\leqslant 2(V_c(n, \phi)) \\ &\leqslant 2(V_c(n, \phi) + \varepsilon) \ . \end{aligned} \tag{6.26}$$

On the other hand

$$\begin{aligned} \sum_{i=1}^{n+1} \int_M (X_i \circ g \circ \phi)^2 dV &= \int_M \sum_{i+1}^{n+1} (X_i \circ g \circ \phi)^2 dV \\ &= V(M) \ , \end{aligned} \tag{6.27}$$

since

$$\Sigma X_i^2 = 1 \ .$$

Using equation (6.21) we find

$$\lambda_1 \Sigma \int_M (X_i \circ g \circ \phi)^2 dV \leqslant \sum_i \int_M |X_i \circ g \circ \phi|^2 dV \ ,$$

which implies

$$\lambda_1 V(M) \leqslant 2(V_c(n, M) + \varepsilon) \ .$$

We now let $\varepsilon \to 0$ and get the inequality

$$\lambda_1 V(M) \leqslant 2V_c(n, M) \ . \tag{6.28}$$

Summarising our results so far, we have proved

Theorem 53. *Let M be a compact 2-dimensional Riemannian manifold and let $\lambda_1 > 0$ be the first non-zero eigenvalue for the Laplacian. We assume that there exists a conformal mapping $\phi : M \to S^n$ so that V_c (n, M) is defined. Then we have the inequality*

$$\lambda_1 V(M) \leqslant 2V_c(n, M) \ .$$

We pass on to the more difficult problem of considering what happens when the above inequality becomes an equality. In particular we shall prove

Theorem 54. *If equality holds in (6.28), then M must be a minimal surface of S^n. Moreover, the immersion is given by eigenfunctions belonging to the first eigenspace.*

Again we follow closely the paper by Li and Yau. By introducing a suitable scaling factor we can assume that $\lambda_1 = 2$, so that our assumption can be written

$$V(M) = V_c(n, M) \ . \tag{6.29}$$

Let $\phi_k : M \to S^n$ be a sequence of conformal mappings such that

$$\lim_{k \to \infty} V_c(n, \phi_k) = V_c(n, M) \ , \tag{6.30}$$

and

$$\int_M X_i \circ \phi_k dV = 0 \tag{6.31}$$

for all i and k. Further, we may assume that the X_i have been chosen so that

$$\lim_{k \to \infty} \int_M (X_i \circ \phi_k)^2 dV > 0 \tag{6.32}$$

for $1 \leqslant i \leqslant N$ for some integer N, and

$$\lim_{k\to\infty} \int_M (X_i \circ \phi_k)^2 dV = 0 \tag{6.33}$$

for $N+1 \leqslant i < n+1$.

Equations (6.21), (6.26), with $\lambda_1 = 2$ imply

$$2V_c(n, \phi_k) \geqslant \sum_{i=1}^{n+1} \int_M |\nabla X_i \circ \phi_k|^2 dV$$

$$\geqslant 2 \sum_{i=1}^{n+1} \int_M (X_i \circ \phi_k)^2 \ . \tag{6.34}$$

Taking the limit as $k \to \infty$ gives

$$V_c(n, \phi_k) \geqslant \lim_{k\to\infty} \sum_{i=1}^{n+1} \int_M (X_i \circ \phi_k)^2 dV \ . \tag{6.35}$$

However since

$$\sum_{i=1}^{n+1} \int_M (X_i \circ \phi_k)^2 dV = V(M) \ , \tag{6.36}$$

we have

$$\lim_{k\to\infty} \sum_{i=1}^{n+1} \int_M (X_i \circ \phi_k)^2 dV = V(M) \ . \tag{6.37}$$

Using (6.32) and (6.33) we get

$$\lim_{k\to\infty} \sum_{n=1}^{N} \int_M (X_i \circ \phi_k)^2 dV = V(M) \tag{6.38}$$

Assumption (6.29) implies that (6.34) become equalities when $k \to \infty$. We can assume that the functions $X_i \circ \phi_k$ converge to some function ψ_i. Moreover

$$\sum_{i=1}^{n+1} \psi_i^2 = 1 \tag{6.39}$$

almost everywhere, and

$$\psi_i = 0 \tag{6.40}$$

for $N+1 \leqslant i \leqslant n+1$.

Also we have

$$\lim_{k\to\infty} \int_M |\nabla X_i \circ \phi_k|^2 dV = \lambda_1 \int \psi_i^2 dV$$

for $1 \leqslant i \leqslant N$.

Thus the sequences $\{X_i \circ \phi_k\}$ converge to $\{\psi_i\}$ and ψ_i are first eigenfunctions of M. Thus $\psi_1, \ldots, \psi_N$ define a smooth conformal mapping of M into S^{N-1}.

We take the Laplacian of (6.39) to get

$$\sum_{i=1}^{N} |\nabla \psi_i|^2 = 2\sum_{1}^{N} \psi_i^2 = 2 = \lambda_1 \ .$$

We already know that $(\psi_1, \ldots, \psi_N)$ is conformal, so we deduce that this gives an isometry. Thus M is mapped as a minimal submanifold of S^{N-1} given by $(\psi_1, \ldots, \psi_N)$. This completes the proof of the theorem.

Corollary 55. *Let M be a compact surface of genus g. Then*

$$\lambda_1 V(M) \leqslant 8\pi(1+g) \ . \tag{6.41}$$

From theorem 53 it will be sufficient to estimate an upper bound of the conformal volume of M. By the Riemann-Roch Theorem, we know that there exists a map $\phi : M \to S^2$ which is conformal and has the property that $|\deg(\phi)| \leqslant 2(1+g)$. Hence, by theorem 51 we have

$$\begin{aligned} V_c(2, M) &\leqslant 2(1+g) V_c(2, S^2) \\ &= 8\pi(1+g) \ , \end{aligned}$$

which yields the result.

Corollary 56. *Let T^2 be a compact surface of genus 1. If T^2 is conformally equivalent to a flat torus with lattice generated by $\{1, 0), (x, y)\}$ where $-\frac{1}{2} \leqslant x \leqslant \frac{1}{2}$ and $\sqrt{(1-x^2)} \leqslant y \leqslant 1$, then*

$$2\pi^2 \leqslant V_c(M) \ . \tag{6.42}$$

To prove the corollary, by theorem 53, it is sufficient to prove that

$$4\pi^2 < \lambda_1(T^2) V(T^2) \ , \tag{6.43}$$

where λ_1 (T^2) and $V(T^2)$ are computed with respect to the flat metric. The area of the fundamental parallelogram of the lattice is just $1.y = y$. Thus

$$V(T^2) = y \ .$$

We have already proved from (6.20) that

$$\lambda_1(T^2) = \frac{4\pi^2}{y^2} \ , \quad \text{when } \sqrt{(1-x^2)} \leqslant y \leqslant 1 \ .$$

Thus we have

$$\lambda_1(T^2)V(T^2)=\frac{4\pi^2}{y}\geqslant 4\pi^2 ,$$

and corollary 56 is proved. Note that the equality sign holds only for the "square" torus generated by the unit square.

6.4 MINIMAL SURFACES AND MINIMAL SUBMANIFOLDS OF UNIT SPHERES

We first prove

Theorem 57. *Let M^2 be a compact minimal surface of S^n given by the immersion $\phi: M^2 \to S^n$. Then*

$$V_c(n, \phi) = V(M) . \tag{6.44}$$

To prove the theorem, let $\pi: S^n \to R^n$ denote stereographic projection. Then since π is a conformal map, it follows that the composite map $\pi \circ \phi$ is a conformal map of M into R^n. Let ν^α be a unit normal vector of M in R^n, and let $\{\mu_i^\alpha\}$, $i=1, 2$ be the principal curvatures associated with ν^α. Then if (ν^α) is a basis for the unit normal space of M, we know from chapter 5 that

$$\int_{\pi \circ \phi(M)} \sum_\alpha (\mu_1^\alpha - \mu_2^\alpha)^2 dV$$

is invariant under a conformal change of metric on R^n. We thus have

$$\int_{\pi \circ \phi(M)} \sum_\alpha (\mu_1^\alpha - \mu_2^\alpha)^2 dV = \int_{g \circ \phi(M)} \sum_\alpha (\bar{\mu}_1^\alpha - \bar{\mu}_2^\alpha)^2 \tag{6.45}$$

where the $\bar{\mu}_i^\alpha$'s are the corresponding principal curvatures of $g \circ \phi(M)$ in S^n.

However, the Gauss equation for submanifolds gives

$$4 \int_{\pi \circ \phi(M)} (|H|^2 - K) dV = 4 \int_{\pi \circ \phi(M)} (|\bar{H}^2| - \bar{K}) d\bar{V} + 4V(g \circ \phi(M)) . \tag{6.46}$$

However, the Gauss-Bonnet theorem enables the integrals involving K and $\bar{K}$ to be cancelled giving

$$\int_{\pi \circ \phi(M)} |H^2| dV = \int_{g \circ \phi(M)} |\bar{H}|^2 dV_g + V(g \circ \phi(M)) . \tag{6.47}$$

The left hand side of this equation is independent of g. So we can deduce, making use of the assumption that $\phi(M)$ is minimal, that

$$\begin{aligned} V(\phi(M)) &= \int_{\pi \circ \phi(M)} |H^2| dV \\ &= \int_{g \circ \phi(M)} |\bar{H}|^2 dV_g + V(g \circ \phi(M)) \\ &\geqslant V(g \circ \phi(M)) . \end{aligned}$$

This gives the required inequality

$$V_c(n, \phi) \leqslant V(M) \ , \tag{6.48}$$

and theorem 57 is proved.

As we have seen, the volume function corresponding to a given immersion $\phi: M \rightarrow S^n$ depends upon $g \in G$ according to the formula

$$V(n, \phi, g) = \int_M dV_g \ ,$$

where dV_g is the volume element corresponding to the metric $\phi^* g^* ds_0^2$. Since S^n is invariant under the orthogonal group $O(n+1)$ acting on E^{n+1}, we have $O(n+1) \subset G$. Let $h \in O(n+1)$. Then

$$V(n, \phi, g) = V(n, \phi, gh) \ ,$$

and hence the volume function can be regarded as defined on the homogeneous coset space $G/O(n+1)$. By an abuse of notation we denote $g\,O(n+1)$ by g. Then we have

$$V_c(n, \phi) = \sup_{g \in G/O(n+1)} V(n, \phi, g) \ . \tag{6.49}$$

We now state without proof the following theorem, and refer the reader to the Li-Yau paper, page 16, for a detailed proof.

Theorem 58. *Let M be a homogeneous Riemannian manifold of dimension m. Let $\phi: M \rightarrow S^n$ be an immersion of M into S^n which has the following properties:*

(i) ϕ is an isometric minimal immersion;
(ii) the transitive subgroup H of the isometry group of M is induced by a subgroup of $O(n+1)$ which we still denote by H;
(iii) $\phi(M)$ does not lie in any hyperplane of R^{n+1}, that is, ϕ is "substantial". Then it follows that

$$V_c(n, \phi) = V(M) \ .$$

Indeed, the identity element of $G/O(n+1)$ is the only local maximum for the volume function defined on $G/O(n+1)$.

We now consider three important corollaries which follow from this theorem.

Corollary 59. *Let M be a compact minimal surface immersed in S^n and let ds^2 be the induced metric. Let $d\bar{s}^2$ be any metric which is conformally equivalent to ds^2, and let $\bar{\lambda}_1$ and $\bar{V}(M)$ denote the first eigenvalue and the volume computed for the metric $d\bar{s}^2$. We assume that the minimal immersion ϕ is given by a subspace of the first eigenspace so that $\bar{\lambda}_1 = 2$. Then we have*

$$V_c(M) = V_c(n, M) = V_c(n, \phi) = \bar{V}(M) \ .$$

To prove the corollary we note that the first equality follows from theorems 53, 54 and 57. Using the hypothesis $\bar{\lambda}_1 = 2$, we get

$$2\bar{V}(M) \leqslant 2V_c(M) \leqslant 2V_c(n, M)$$

by theorem 53. But we know that

$$\begin{aligned} V_c(n, M) &\leqslant V_c(n, \phi) \\ &= \bar{V}(M) \ . \end{aligned}$$

Hence all four terms in the theorem must be equal.

When M is the 2-sphere, S^2, since the conformal structure is unique, it follows that for any metric $\bar{d}s^2$ on S^2 we have

$$\bar{\lambda}_1 \bar{V}(S^2) \leqslant 8\pi = 2V_c(S^2) \ . \tag{6.50}$$

This result was previously proved by J. Hersch in 1970.

Corollary 60. *Let $\bar{d}s^2$ be any metric on the real projective plane $P^2(\mathbb{R})$. Then*

$$\lambda_1 V P^2(\mathbb{R}) \leqslant 2V_c P^2(\mathbb{R}) = 12\pi \ .$$

If equality holds, then there exists a subspace of the first eigenspace of the Laplacian for $d\bar{s}^2$ which gives an isometric minimal immersion of $P^2(\mathbb{R})$ into S^4.

To prove the corollary, we observe that, like the sphere, $P^2(\mathbb{R})$ has a unique conformal structure. We have seen that the first eigenspace of $P^2(\mathbb{R})$ with respect to the standard metric gives an isometric minimal embedding of $P^2(\mathbb{R})$ into S^4. This is the Veronese surface whose volume we have calculated to be 6π. The corollary now follows from theorems 53, 54 and 57.

The flat square torus can be isometrically minimally immersed into S^3 by

$$y^1 = \frac{1}{\sqrt{2}} \cos u \ , \quad y^2 = \frac{1}{\sqrt{2}} \sin u \ , \quad y^3 = \frac{1}{\sqrt{2}} \cos v \ , \quad y^4 = \frac{1}{\sqrt{2}} \sin v$$

where $\quad 0 \leqslant u \leqslant 2\pi \ , \quad 0 \leqslant v \leqslant 2\pi \ ,$

the functions being eigenfunctions of the first eigenvalue. So we can derive the following:

Corollary 61. *Let M be a compact surface of genus 1. Let the Riemannian metric $\bar{d}s^2$ on M be conformally equivalent to the square torus with lattice generated by (1, 0) and (0, 1). Then*

$$\bar{\lambda}_1 \bar{V}(M) \leqslant 4\pi^2.$$

If equality holds, then M can be isometrically minimally immersed into S^3 by functions of the first eigenspace of the Laplacian.

6.5 THE WILLMORE CONJECTURE AND CONFORMAL AREA

It will be remembered that in chapter 5 we referred to the conjecture that

$$\int_{T^2} |H|^2 dS \geqslant 2\pi^2$$

for any immersed torus T^2 in R^3.

We shall now see how the theory of conformal area developed by Li and Yau sheds considerable light on this and related problems. We consider the more general problem where M is a compact surface immersed in any higher dimensional euclidean space R^n. The essential observation is the following:

Theorem 62. *Let M be a compact surface immersed in R^n. Then*

$$\int_M |H|^2 dS \geqslant V_c(n, M) \ . \tag{6.51}$$

Moreover, equality implies that M is the image of some minimal surface in S^n under a suitable stereographic projection.

We compose the above immersion of M into R^n with the inverse of stereographic projection to get a conformal immersion ϕ of M into S^n. The area of $\phi(M)$ will in general differ from the n-conformal area of ϕ. However, if we compose this with a suitable Möbius transformation, we can ensure that this area is, in fact, equal to $V_c(n, \phi)$.

Using the same argument as in the proof of Theorem 57 we have

$$\int_M |H^2| dS = \int_{\phi(M)} |\bar{H}|^2 dS^n + V(\phi(M)) \ , \tag{6.52}$$

where $\bar{H}$ is the mean curvature vector of $\phi(M)$ in S^n. Since we have assumed that $V(\phi(M)) = V_c(n, \phi)$ we have the inequality

$$\int_M |H|^2 dS \geqslant V_c(n, \phi)$$

which is required.

We can now deduce

Theorem 63. *Let M be a compact surface in R^n. Then*

$$\int_M |H^2| \mathrm{d}S \geqslant \tfrac{1}{2} \sup \{\lambda_1 V(M)\} \tag{6.53}$$

where the supremum is taken over any metric which is conformally equivalent to the induced metric from R^n.

Moreover, as consequences of corollaries 56 and 60 we have:

Theorem 64. *Let M be a compact surface homeomorphic to $P^2(\mathbb{R})$. Then for any immersion of M into R^n we have*

$$\int_M |H^2| dS \geqslant 6\pi \ . \tag{6.54}$$

Equality implies that M is the image of a stereographic projection of some minimal surface in S^4 with $\lambda_1 = 2$.

Theorem 65. *Let M be a surface of genus 1 in R^n. Suppose M is conformally equivalent to one of the flat tori described in corollary 56. Then*

$$\int_M |H^2| dS \geqslant 2\pi^2 \ . \tag{6.55}$$

Equality implies that M must be conformally equivalent to the square torus and that it is the image of a stereographic projection of a minimal torus in S^3.

To prove the above theorem, we note that the first part follows from corollary 56 and theorem 62. When equality holds, the proof of corollary 56 implies that M is conformally equivalent to a flat torus with lattice generated by $(1, 0)$ and $(x, 1)$. We know, however, that the space of eigenfunctions corresponding to λ_1 must give an isometric minimal immersion into S^3 and this forces the relation $x = 0$. The last part of the theorem follows as a consequence of theorem 62.

An interesting exercise is to try to generalize theorem 65 to surfaces of genus > 1. In particular the following conjecture of Li and Yau seems plausible:

Conjecture 66. *If M can be conformally embedded as a minimal surface in S^3, then $\int_M H^2 dS$ is not less than the area of this minimal surface.*

So far this conjecture remains an open question.
However, we prove

Theorem 67. *Let $\psi: M \to R^n$ be an immersion of a compact surface. Let p be a point of R^n such that $\psi^{-1}(p) = \{x_i\}$, $i = 1, 2, \ldots, k$ where the k points x_i of R^n are all distinct. Then*

$$\int_M H^2 dS \geqslant 4k\pi \ . \tag{6.56}$$

To prove this theorem, let $\pi: S^n \to R^n$ denote a stereographic projection.

Then the composite map $\pi^{-1}\circ\psi$ is a conformal map of M into S^n. As in the proof of theorem 57 we have

$$\int_M |H|^2 dS \geqslant V_c(n, \pi^{-1}\circ\psi) \ , \tag{6.57}$$

so it is sufficient to compute a lower bound of $V_c(n, \pi^{-1}\circ\psi)$.

If we use the "blowing up" technique used in Theorem 52 at the point $\pi^{-1}\circ\psi(x_i)$, we get k copies of S^2, one from the tangent plane at each point x_i. Thus we have

$$V_c(n, \pi^{-1}\circ\psi) \geqslant kV(S^2) = 4k\pi \tag{6.58}$$

as claimed.

Corollary 68. *Let ψ: $M \to R^n$: have the property that*

$$\int_M |H|^2 dS < 8\pi \ . \tag{6.59}$$

Then ψ must be an embedding.

This is an immediate consequence of Theorem 67. We also have the following theorems which are consequences of (6.52).

Theorem 69. *Let M be a compact surface homeomorphic to $P^2(\mathbb{R})$. If M is a minimum surface of some unit sphere S^n, then*

$$V(M) \geqslant 6\pi = \text{volume of the Veronese surface.} \tag{6.60}$$

Theorem 70. *Let M be a compact surface of genus 1. Suppose M is conformally equivalent to one of the tori described in corollary 56, and that M is a minimal surface in some unit sphere S^n. Then*

$$V(M) \geqslant 2\pi^2 \ . \tag{6.61}$$

Equality implies that M is conformally equivalent to the square torus.

Theorem 71. *Let ϕ: $M \to S^n$ be a minimal immersion of a compact surface M into some unit sphere S^n. Let p be a point of S^n such that $\phi^{-1}(p)$ consists of k distinct points of M. Then*

$$V(M) \geqslant 4k\pi \ . \tag{6.62}$$

In particular if $V(M) < 8\pi$, then ϕ must be a minimal embedding.

6.6 HARMONIC MAPS

In this section we describe a differential geometric invariant which generalizes the concept of mean curvature vector. This was first introduced in a fundamental research paper by Eells and Sampson (1964), and has developed into a substantial branch of pure mathematics with considerable applications to physics. The best general survey available is Eells and Sampson (1978). We give below an outline of the theory which is most relevant to our purpose.

Let (M, g), (N, h) be Riemannian manifolds with smooth metrics g and h. Let $\phi:(M, g)\to(N, h)$ be a smooth map. The pull-back ϕ^*h is a symmetric semi-definite covariant tensor field of the second order which we call the **first fundamental form** of ϕ. This is effectively the induced "metric" on M. The bundle over M pulled back from the tangent bundle TN over N has a Riemannian structure carried over from that of $TN\to N$. The differential $d\phi$ of ϕ can be interpreted as a $\phi^{-1}T(N)$-valued 1-form on M; that is

$$d\phi\in\mathscr{C}(T^*(M)\otimes\phi^{-1}T(N)) \; ,$$

where the right-hand side denotes the class of $\phi^{-1}T(N)$-valued 1-forms on M.

The exterior derivative of $d\phi$ given by $d(d\phi)$ is clearly zero and so gives nothing new. On the other hand, the covariant differential $\nabla(d\phi)$ is a non-trivial symmetric second order covariant tensor with values in $\phi^{-1}T(N)$. We call this the **second fundamental form** of the map ϕ.

Let $U\subset M$ be a domain with coordinates $(x^1, \ldots, x^m)$ and let $V\subset N$ be a domain with coordinates $(y^1, \ldots, y^n)$ such that $\phi(U)\subset V$. Locally the map ϕ is represented by

$$y^\alpha=\phi^\alpha(x^1, \ldots, x^m) \; , \quad \alpha=1, \ldots, n \; .$$

The metric tensor g is represented by $g(x)=g_{ij}(x)dx^i dx^j$ (using summation convention), where $i, j=1, \ldots, n$. Similarly h is represented by $h(y)=h_{\alpha\beta}(y)dy^\alpha dy^\beta$. The differential $d\phi(x)$ has matrix representation $(\partial\phi^\alpha/\partial x^i)$. The first and second fundamental forms of ϕ at $x\in U$ can be represented by

$$(\phi^*h)_{ij}=\frac{\partial\phi^\alpha}{\partial x^i}\frac{\partial\phi^\beta}{\partial x^j}h_{\alpha\beta} \tag{6.63}$$

$$(\nabla(d\phi))^\gamma_{ij}=\frac{\partial^2\phi^\gamma}{\partial x^i\partial x^j}-{}^M\Gamma^k_{ij}\frac{\partial\phi^\gamma}{\partial x^k}+{}^N\Gamma^\gamma_{\alpha\beta}\frac{\partial\phi^\alpha}{\partial x^i}\frac{\partial\phi^\beta}{\partial x^j} \; , \tag{6.64}$$

where ${}^M\Gamma$ and ${}^N\Gamma$ are the Christoffel symbols of the Levi-Civita connexions of g and h.

The trace of the second fundamental form is denoted by $\tau(\phi)\in\mathscr{C}(\phi^{-1}T(N))$ and is called the **tension field** of ϕ. A map ϕ is called **harmonic** if and only if $\tau(\phi)=0$. In terms of coordinates we get

$$\tau^\gamma(\phi)=g^{ij}(\nabla(d\phi))^\gamma_{ij} \; . \tag{6.65}$$

Following Eells and Lemaire, we can get a physical picture of a harmonic map $\phi: M \to N$ as follows. We suppose that M is made of "rubber" and that N is made of "marble": the map ϕ constrains M to lie on N. At each point $x \in M$ we have a vector $\tau(x)$ representing the tension in the rubber at that point. We see that ϕ is harmonic if and only if ϕ constrains M to lie on N in a position of elastic equilibrium. This explains the terminology "tension" field.

Another way in which the tension field arises is as follows. We suppose that M is compact. We define the *energy* of ϕ at $x \in M$ by the function

$$e(\phi)(x) = \tfrac{1}{2}\,\text{Trace}\,(\phi^* h)(x) \ . \tag{6.66}$$

Thus the energy of ϕ at x is equal to one half of the sum of the eigenvalues of the first fundamental form $(\phi^* h)(x)$ on $T_x(M)$ with respect to $g(x)$.

The energy of $\phi: M \to N$ is defined by

$$E(\phi, M) = \int_M e(\phi)(x) dV \ . \tag{6.67}$$

In terms of coordinates, we have

$$e(\phi)(x) = \tfrac{1}{2} g^{ij} \frac{\partial \phi^\gamma}{\partial x^i} \frac{\partial \phi^\beta}{\partial x^j} h_{\alpha\beta} \ . \tag{6.68}$$

In the first paper on the subject, Eells and Sampson defined a map ϕ to be harmonic if and only if it is an extremal of the energy integral (6.67).

The Euler-Lagrange equation arising from the variational problem of the energy integral can be shown to be

$$\tau(\phi) = 0 \ . \tag{6.69}$$

This relation shows the compatibility of the two definitions of harmonic maps.

Harmonic maps appear in very many different contexts. For example:

(a) If $\dim M = 1$, then harmonic maps are geodesics of N.
(b) If $N = \mathbb{R}$, they are the harmonic functions of M.
(c) If $\dim M = 2$, they include the minimal surfaces of N.
(d) If M is a Riemannian submanifold of N of minimal volume, then the inclusion map $i: M \to N$ is harmonic.

We see that the tension field τ is a natural generalization of the mean curvature vector field, and effectively coincides with it when the map ϕ is an isometry.

This suggests the problem of calculating

$$\int_M |\tau|^2 dS \text{ instead of } \int_M |H|^2 dS \ .$$

It seems probable that the whole theory of conformal volume discussed earlier in this chapter could be modified to deal with maps $\phi : (M, g) \to (N, h)$, this time replacing the group of isometries of S^n by the group of isometries of (N, h). However, we leave the reader to investigate further this exciting field of research.

We note that for an isometric immersion, $e(\phi) = m/2$. It follows that if ϕ is an isometry, the extremals of the energy integral are precisely the minimal submanifolds of N. Alternatively we may say that the mean curvature vector associated with a harmonic isometric immersion is zero.

More generally, suppose that $\phi : M \to N$ is an isometric immersion so that $\phi^* h = g$. Let $V(N, M) \to M$ denote the normal bundle of M in N. Then $d\phi$ is a 1-form on M with values in $\phi^{-1}T(N)$, and its derivative $\beta = \nabla(d\phi)$ is just the *second fundamental form of the immersion*. The mean curvature vector field H is given by Trace β/m.

In this case, it is easy to verify that the tension field $\tau(\phi)$ is m-times the mean curvature vector H. *Thus ϕ is harmonic if and only if ϕ is minimal.*

Bibliography

This bibliography contains quite a number of references in addition to those cited in the text, primarily those concerned with 'taut immersions' and 'isoparametric surfaces'.

Avez, A. (1963), Applications de la formule de Gauss-Bonnet-Chern aux variétés à quatre dimensions, *C.R.Acad.Sci.*, **256,** 5488–5490.

Banchoff, T. F. (1965), Tightly embedded 2-dimensional polyhedral manifolds, *Amer. J. Math.*, **87,** 462–472.

Banchoff, T. F. (1967), Critical points and curvature for embedded polyhedra, *J. Diff. Geom.* **1,** 245–256.

Banchoff, T. F. (1970), The spherical two-piece property and tight surfaces in spheres, *J. Diff. Geom.*, **4,** 193–205.

Banchoff, T. F. (1971), The two-piece property and tight n-manifolds-with-boundary in E^n, *Trans. Amer. Math. Soc.* **161,** 259–267.

Banchoff, T. F. (1971), High codimensional 0-tight maps on spheres *Proc. Amer. Math. Soc.*, **29,** 133–135.

Banchoff, T. F. (1974), Tight polyhedral Klein bottles, projective planes and Moebius bands, *Math. Ann.*, **207,** 233–243.

Berger, M., Gauduchon, P. and Mazet, E. (1971), Le spectre d'une variété Riemannienne, *Lecture Notes in Math.*, **194,** Springer-Verlag.

Bishop, R. L. and Goldberg, S. I. (1964), Some implications of the generalized Gauss-Bonnet Theorem, *Trans. Amer. Math. Soc.*, **112,** 508–535. MR29 # 574.

Blaschke, W. (1929), Vorlesungen über Differentialgeometrie III, Springer, Berlin.

Borsuk, K. (1947), Sur la courbure totale des courbes, *Ann. de la Soc. Math. Pol.*, **20,** 251–265.

Bourguignon, J. P. and Karcher, H. (1978), Curvature operators: Pinching estimates and geometric examples, *Ann. scient. Éc. Norm. Sup.*, **11,** 71–92.

Cartan, É. (1938), Familles de surfaces isoparametriques dans les espaces à courbure constante, *Annali di Mat.*, **17,** 177–191.

Cartan, É. (1939), Sur des familles remarquables d'hypersurfaces isoparametriques dans les espaces spheriques, *Math. Z.*, **45,** 335–367.

Cartan, É. (1939), Sur quelques familles remarquables d'hypersurfaces, *C.R. Congres. Mat. Liege*, 30–41.

Cartan, É. (1940), Sur des familles remarquables d'hypersurfaces isoparametriques des espaces spheriques à 5 et 9 dimensions, *Revista Univ. Tucuman, Serie A*, **1,** 5–22.

Carter, S. and West, A. (1972), Tight and taut immersions, *Proc. Lond. Math. Soc.*, **25,** 701–720.

Carter, S. and West, A. (1978), Totally focal embeddings, *J. Diff. Geom.* **13,** 251–261.

Carter, S. and West, A. (1981), A characterization of isoparametric hypersurfaces in spheres, preprint.

Cecil, T. (1974), Geometrical applications of critical point theory to submanifolds of complex projective space, *Nagoya Math. J.*, **55,** 5–31.

Cecil, T. (1974), A characterization of metric spheres in hyperbolic space by Morse theory, *Tohoku Math. J.*, **16,** 341–351.

Cecil, T. (1976), Taut immersions of non-compact surfaces into a Euclidean 3–space, *J. Diff. Geom.*, **11,** 451–459.

Cecil, T. E. and Ryan, P. J. (1978), Focal sets of submanifolds, *Pacific J. Math.*, **78,** 27–39.

Cecil, T. E. and Ryan, P. J. (1978), Focal sets, taut embeddings and the cyclides of Dupin, *Math. Ann.*, **236,** 177–190.

Cecil, T. E. and Ryan, P. J. (1979), Distance functions and umbilic submanifolds of hyperbolic space, *Nagoya Math. J.*, **74,** 67–75.

Cecil, T. E. and Ryan, P. J. (1979), Tight and taut immersions into hyperbolic space, *J. Lond. Math. Soc.*, **19,** 561–572.

Cecil, T. E. and Ryan, P. J. (1980), Conformal geometry and the cyclides of Dupin, *Canadian J. Math.*, **32,** 767–782.

Cecil, T. E. and Ryan, P. J. (1981), Tight spherical embeddings, Proceedings, Berlin, Symposium in Global Differential Geometry, *Lectures Notes in Mathematics*, **838**, Springer-Verlag, Berlin, Heidelberg, New York, 94–104.

Cecil, T. E. and Ryan, P. J. (1982), Focal sets and real hypersurfaces in complex projective space, *Trans. Amer. Math. Soc.*, **269,** 481–499.

Chen, B.-Y. (1971), On the total curvature of immersed manifolds I, *Amer. J. Math.*, **93,** 148–162.

Chen, B.-Y. (1972), On the total curvature of immersed manifolds II, *Amer. J. Math.*, **94,** 799–809.

Chen, B.-Y. (1973), An invariant of conformal mappings, *Proc. Amer. Math. Soc.* **40,** 563–564.

Chen, B.-Y. (1973), Geometry of Submanifolds, Marcel Dekker.

Chen, B.-Y. (1973), On a variational problem on hypersurfaces, *J. Lond. Math. Soc.* (2), **6,** 321–325.

Chen, B.-Y. (1973), On the total curvature of immersed manifolds III, *Amer. J. Math.*, **95,** 636–642.

Chen. B.-Y. (1974), Some conformal invariants of submanifolds and their applications, *Bol. Un. Mat. Ital.* (*4*) **10,** 380–385.

Chen, B.-Y. (1976), Total mean curvature of immersed surfaces in E^m, *Trans. Amer. Math. Soc.*, **216,** 333–341.

Chen, B.-Y. (1979), On the total curvature of immersed manifolds IV, *Bull.*

Inst. Math. Acad. Sinica, **7,** 301–311.
Chen, B.-Y. (1979), Conformal mappings and first eigenvalues of Laplacian on surfaces, *Bull. Inst. Math. Acad. Sinica*, **7,** 395–400.
Chen, B.-Y. (1981), On the total curvature of immersed manifolds V, *Bull. Inst. Math. Acad. Sinica*, **9,** 509–516.
Chen, C. S. (1972), On tight isometric immersions of codimension two, *Amer. J. of Math.*, **94,** 974–990.
Chen, C. S. (1973), More on tight isometric immersions of codimension two, *Proc. Amer. Math. Soc.*, **40,** 545–553.
Cheng, S. Y. (1976), Eigenfunctions and nodal sets, *Comm. Math. Helv.*, **51,** 43–55.
Cheng, S. Y., Li, P. and Yau, S. T. (1982), Heat equation on minimal submanifolds and their applications, to appear in *Am. J. Maths.*
Chern, S. S. (1955), La géomètrie des sous-variétés d'un espace euclidien à plusieurs dimensions, *Enseignement Math.*, **40,** 26–46.
Chern, S. S. (1956), On curvature and characteristic classes of a Riemannian manifold, *Abh. Math. Sem. Univ. Hamburg*, **20,** 117–126; MR 17 # 783.
Chern, S. S. (1967), Curves and surfaces in euclidean spaces, Studies in global geometry and analysis, MAA Studies in Mathematics, **4,** 26–56.
Chern, S. S. and Lashof, R. K. (1957), On the total curvature of immersed manifolds I, *Amer. Journ. Math.*, **79,** 306–318.
Chern, S. S. and Lashof, R. K. (1958), On the total curvature of immersed manifolds, II, *Mich. Math. Journal*, **5,** 5–12.
Chern, S. S. (1968), Minimal Submanifolds in a Riemannian Manifold, University of Kansas, Technical Report 19.
Courant, R. and Hilbert, D. (1962), Methods of Mathematical Physics, Interscience.
Eells, J. and Kuiper, N. (1962), Manifolds which are like projective planes, *Institut des Hautes Etudes Sci. Pub. Maths.*, **14,** 5–46.
Eells, J. and Sampson, J. H. (1964), Harmonic mappings of Riemannian manifolds, *Amer. J. Math.*, **86,** 109–160.
Eells, J. and Lemaire, L. (1978), A report on harmonic maps, *Bull. Lond. Math. Soc.*, **10,** 1–68.
Fary, I. (1949), Sur la courbure totale d'une courbe gauche faisant un noeud, *Bull. Soc. Math. France*, **77,** 128–138.
Faux, I. D. and Pratt, M. J. (1979), Computational Geometry for Design and Manufacture, Ellis Horwood.
Fenchel, W. (1929), Über die Krümmung und Windung geschlossenen Raumkurven, *Math. Ann.*, **101,** 238–252.
Ferus, D. (1968), Totale Absolutkrümmung in Differentialgeometrie und Topologie, *Lecture Notes in Mathematics*, **66,** Springer-Verlag, Berlin, Heidelberg, New York.
Ferus, D. (1971), Über der absolute Totalkrümmung höher-dimensionaler Knoten, *Math. Ann.*, **171,** 81–86.
Ferus, D. (1980), Symmetric submanifolds of euclidean space, *Math. Ann.*, **247,** 81–93.
Ferus, D., Karcher, H. and Munzner, H. F. (1981), Cliffordalgebren und neue

isoparametrischen Hyperflachen, *Math. Z.*, **177,** 479–502.

Fox, R. H. (1950), On the total curvature of some tame knots, *Ann. Math.* **52,** 258–261.

Freudenthal, H. (1953), Zur ebenen Oktavengeometrie, Proc. Akad. Amsterdam A 56; *Indag. Math.*, **15,** 195–200.

Freudenthal, H. (1951), Octaven, Ausnahmegruppen und Oktavengeometrie, Math. Inst. Univ. Utrecht, Preprint.

Geroch, R. (1974), Positive sectional curvature does not imply positive Gauss-Bonnet Integrand, *Proc. A. M. S.*, **54,** 267–270.

Gray, A. and Vanhecke, L. (1979), Riemannian geometry as determined by the volume of small geodesic balls, *Acta Math.*, **142,** 157–198.

Hebda, J. (1981), Manifolds admitting taut hyperspheres, *Pacific J. Math.*, **97,** 119–124.

Helgason, S. (1978), Differential Geometry, Lie Groups and Symmetric Spaces, Academic Press.

Hersch, J. (1970), Quatre propriétés isopérimétriques de membranes sphériques homogènes, *C. R. Acad. Sci.*, **270,** 1645–1648.

Hurwitz, A. (1923) Über die Komposition der quadratischen Formen, *Math. Ann.*, **88,** 1–25.

Husemöller, (1966), Fibre Bundles, McGraw Hill.

Kelly, E. F. (1971), 0-tight equivariant imbeddings of symmetric spaces, *Bull. Amer. Math. Soc.*, **77,** 580–583.

Kelly, E. F. (1972), Tight equivariant imbeddings of symmetric spaces, *J. Diff. Geom.*, **7,** 535–548.

Klembeck, P. (1976), On Geroch's counter-example to the algebraic Hopf conjecture, *Proc. A. M. S.*, **59,** 334–336.

Kobayashi, S. (1967), Imbeddings of homogeneous spaces with minimum total curvature, *Tohoku Math. J.*, **19,** 63–70.

Kobayashi, S. (1968), Isometric imbeddings of compact symmetric spaces, *Tohoku Math. J.*, **20,** 21–25.

Kobayashi, S. and Nomizu, K. (1963), Foundations in differential geometry, Vol. I, Wiley-Interscience, New York.

Kobayashi, S. and Nomizu, K. (1969), Foundations in differential geometry, Vol. II, Wiley-Interscience, New York.

Kobayashi, S. and Takeuchi, M. (1968), Minimal imbeddings of R-spaces, *J. Diff. Geom.*, **2,** 203–215.

Kon, M. (1979), Pseudo-Einstein real hypersurfaces in complex space forms, *J. Diff. Geom.*, **14,** 339–354.

Kuhnel, W. (1977), Total curvature of manifolds with boundary in E^n, *J. Lond. Math. Soc.* (2), **15,** 173–182.

Kuhnel, W. (1979), Total absolute curvature of polyhedral manifolds with boundary in E^n, *Geom. Ded.*, **8,** 1–12.

Kuhnel, W. (1980), Tight and 0-tight polyhedral embeddings of surfaces, *Invent. Math.*, **58,** 161–177.

Kuiper, N. H. (1959), Immersions with minimal total absolute curvature, Coll. de Géométrie Differentielle Globale (Bruxelles, 1958), *Centre Belg. Rech. Math., Louvain*, 75–88.

Kuiper, N. H. (1959), Sur les immersions à courbure totale minimale, *Sem. de Top. et de Geom. Diff.*, C. Ehresmann, Paris, 1–5.

Kuiper, N. H. (1960), On surfaces of euclidean three-space, *Bull. Soc. Math. de Belgique*, **12,** 5–22.

Kuiper, N. H. (1960), La courbure d'indice k et les applications convexes, Seminaire Ehresmann, 1–15.

Kuiper, N. H. (1961), Convex Immersions of Closed Surfaces in E^3, *Comm. Math. Helvitici*, **35,** 85–92.

Kuiper, N. H. (1962), On convex maps, Nieuw Archief voor Wiskunde, **10,** 147–164.

Kuiper, N. H. (1967), Der Satz von Gauss-Bonnet fur Abbildungen im E^n, *Jahr. Ber. DMV*, **69,** 77–88.

Kuiper, N. H. (1970), Minimal total absolute curvature for immersions, *Invent. Math.*, **10,** 209–238.

Kuiper, N. H. (1971), Morse relations for curvature and tightness, Proc. Liverpool Singularities Symp. II (Ed. by C. T. C. Wall), *Lecture Notes in Mathematics*, Springer, **209**, 77–89.

Kuiper, N. H. (1972), Tight topological embeddings of the Moebius band, *J. Diff. Geom.*, **6,** 271–283.

Kuiper, N. H. (1975), Stable surfaces in Euclidean three-space, *Math. Scand.*, **36,** 83–96.

Kuiper, N. H. (1976), Curvature measures for surfaces in E^n, Lobachevski Colloquium, Kazan, USSR.

Kuiper, N. H. (1980), Tight embeddings and maps. Submanifolds of geometrical class three in E^n, The Chern Symposium 1979, Proc. International Symp., Berkeley, Calif., Springer-Verlag, Berlin Heidelberg, New York, 97–145.

Kuiper, N. H. (1982), Taut sets in three-space are very special, preprint.

Kuiper, N. H. and Meeks, W. H. (1982), Total Absolute curvature for Knotted Surfaces, preprint.

Kuiper, N. H. and Pohl, W. F. (1977), Tight topological embeddings of the real projective plane in E^5, *Invent. Math.*, **42,** 177–199.

Kulkarni, R. S. (1972), On the Bianchi identies, Math. Ann., **199,** 175–204.

Langevin, R. and Rosenberg, H. (1976), On curvature integrals and knots, Topology, **15**, 405–416.

Lawson, H. B. (1968), Minimal varieties in constant curvature manifolds, Ph.D. Thesis, Stanford.

Lawson, H. B. (1970), Rigidity theorems in rank-1 symmetric spaces, *J. Diff. Geom.*, **4,** 349–357.

Lawson, H. B. (1980), Lectures on Minimal Submanifolds, Vol. I, Mathematics Lecture Series **9,** Publish or Perish.

Li, P. and Yau, S. T. (1982), 'A New Conformal Invariant and its Applications to the Willmore Conjecture and the First Eigenvalue of Compact Surfaces', preprint.

Little, J. A. and Pohl, W. F. (1971), On Tight Immersion of Maximal Codimension, *Inventiones Math.*, **13,** 179–204.

Maeda, Y. (1976), On real hypersurfaces of a complex projective space,

J. Math. Soc. Japan, **28,** 529–540.

Matsushima, Y. (1972), Differentiable Manifolds, Marcel Dekker.

Meeks, W. (1979), Lectures on Plateau's problem (July 1978), Escola de Geometria Differencial DO CEARA, Brasil, 1–53,

Milnor, J. W. (1950), On the total curvature of knots, *Ann. Math.*, **52,** 248–257.

Milnor, J. W. (1953), On the total curvature of space curves, *Math. Scand.*, **1,** 289–296.

Milnor, J. W. (1956), On manifolds homeomorphic to the 7-sphere, *Ann. of Math.*, **69,** 399–405.

Milnor, J. W. (1963), Morse Theory, *Ann. Math. Stud.*, **51,** Princeton.

Moore, J. D. (1978), Codimension two submanifolds of positive curvature, *Proc. Amer. Math. Soc.*, **70,** 72–78.

Morse, M. (1960), The existence of polar non-degenerate functions on differentiable manifolds, *Ann. Math.*, **71,** 352–383.

Morton, H. R. (1979), A criterion for an embedded surface in R^3 to be unknotted, Topology of Low-Dimensional Manifolds, Proceedings, Sussex, 1977, *Springer Lectures Notes*, **722,** 93–98.

Munzner, H. F. (1980), Isoparametrische Hyperflachen in Sphären, I, *Math. Ann.*, **251,** 57–71.

Munzner, H. F. (1981), Isoparametrische Hyperflachen in Sphären, II, *Math. Ann.*, **256,** 215–232.

Nomizu, K. (1973), Some results in É. Cartan's theory on isoparametric families of hypersurfaces, *Bull. Amer. Math. Soc.*, **79,** 1184–1188.

Nomizu, K. (1974), Élie Cartan's work on isoparametric families of hypersurfaces, Proc. Symp. in Pure Math., *Amer. Math. Soc.*, **27,** 191–200.

Nomizu, K. and Rodriguez, L. (1972), Umbilical submanifolds and Morse functions, *Nagoya Math. J.*, **48,** 197–201.

Nomizu, K. and Smythe, B. (1968), Differential geometry of complex hypersurfaces, II, *J. Math. Soc. Japan*, **20,** 498–521.

O'Neill, B. (1966), Elementary Differential Geometry, Academic Press.

O'Neill, B. (1966), Fundamental equations of a submersion, *Mich. Math. J.*, **13,** 459–469.

Ozeki, H. and Takeuchi, M. (1975), On some types of isoparametric hypersurfaces in spheres, I, *Tohoku Math. J.*, **27,** 515–559.

Ozeki, H. and Takeuchi, M. (1976), On some types of isoparametric hypersurfaces in spheres, II, *Tohoku Math. J.*, **28,** 7–55.

Palais, R. (1957), A global formulation of the Lie theory of transformation groups, *Mem. Amer. Math. Soc.*, **22.**

Reckziegel, H. (1976), Krümmungsflachen von isometrischen Immersionen in Raume Konstanter Krümmung, *Math. Ann.*, **223,** 169–181.

Reckziegel, H. (1979), On the eigenvalues of the shape operator of an isometric immersion into a space of constant curvature, *Math. Ann.*, **243,** 71–82.

Reckziegel, H. (1979), Completeness of curvature surfaces of an isometric immersion, *J. Diff. Geom.*, **14,** 7–20.

Rodriguez, L. L. (1976), The two-piece property and convexity for surfaces with boundary, *J. Diff. Geom.*, **11,** 235–250.

Rodriguez, L. L. and Guadalupe, I. V. (1981), Normal curvature of surfaces into space forms, preprint.

Ryan, P. J. (1969), Homogeneity and some curvature conditions for hypersurfaces, *Tohoku Math. J.*, **21,** 363–388.

Shiohama, K. and Takagi, A. (1970), A characterization of a standard torus in E^3, *Journ. Diff. Geometry*, **4,** 477–485; M. R. 43 # 2646.

Smyth, B. (1967), Differential geometry of complex hypersurfaces, *Ann. Math.* (2), **85**, 246–266.

Spivak, M. (1970), Differential Geometry, Publish or Perish.

Steenrod, N. A. (1951), The topology of fibre bundles, Princeton Univ. Press.

Szegö, G. (1954), Inequalities for certain eigenvalues of a membrane of given area, *J. Rat. Mech. Anal.*, **3,** 343–356.

Tai, S. S. (1968), On the minimum imbeddings of compact symmetric spaces of rank one, *J. Diff. Geom.*, **2,** 55–66.

Takagi, R. (1975), Real hypersurfaces in a complex projective space with constant principal curvatures, I, *J. Math. Soc. Japan*, **25,** 43–53.

Takagi, R. (1975), Real hypersurfaces in a complex projective space with constant principal curvatures, II, *J. Math. Soc. Japan*, **27,** 507–516.

Takagi, R. (1976), A class of hypersurfaces with constant principal curvatures in a sphere, *J. Diff. Geometry*, **11,** 225–233.

Takagi, R. and Takahashi, T. (1972), On the principal curvatures of homogeneous hypersurfaces in a sphere, Diff. Geom. in honour of K. Yano, Kinokuniya, Tokyo, 469–481.

Teufel, E. (1979), Total Krümmung und totale Absolutkrümmung in der sphärischen Differentialgeometrie und Differentialtopologie, Thesis Stuttgart.

Teufel, E. (1980), Eine Differentialtopologische Berechnung der totalen Krümmung und totalen Absolutkrümmung in der sphärischen Differentialgeometrie, *Manuscripta Math.*, **31,** 119–147.

Teufel, E. (1980), Anwendungen der Differentialtopologischen Berechnung der totalen Absolutkrümmung in der sphärischen Differentialgeometrie, *Manuscripta math.*, **32,** 239–262.

Teufel, E. (1982), Differential Topology and the Computation of Total Absolute Curvature, *Math. Ann.*, **258,** 471–480.

Thomsen, G. (1923), Über konforme Geometrie I: Grundlagen der konformen Flächentheorie, *Abh. Math. Sem. Univ. Hamburg*, 31–56.

Weiner, J. L. (1978), On a problem of Chen, Willmore, et al., *Indiana University Math. Journal*, **27,** 19–35.

Weinstein, A. (1973), Distance Spheres in Complex Projective Spaces, *Proc. Amer. Math. Soc.*, **39,** 649–650.

Weyl, H. (1939), On the volume of tubes, *Amer. J. Math.*, **61,** 461–472.

White, J. H. (1973), A global invariant of conformal mappings in space, *Proc. Amer Math. Soc.*, **38,** 162–164.

Willmore, T. J. (1959), Introduction to Differential Geometry, Oxford, Clarendon.

Willmore, T. J. (1965), Note on Embedded Surfaces, *An. st. Univ. Iasi, s.I.a. Matematica*, **11B,** 493–496.

Willmore, T. J. (1968), Mean curvature of immersed surfaces, *An. st. Univ. Iasi, s.I.a. Matematica*, **14,** 99–103.

Willmore, T. J. (1971), Mean curvature of Riemannian immersions, *Journ. Lond. Math. Soc.* (2), **3,** 307–310.

Willmore, T. J. (1971), Tight immersions and total absolute curvature, *Bull. Lond. Math. Soc.*, **3,** 129–151.

Willmore, T. J. (1974), Minimal conformal immersions, *Lecture Notes in Maths.*, **392**, Springer-Verlag, Berlin, 111–120.

Willmore, T. J. (1976), Mean curvature of immersed manifolds. Topics in Differential Geometry, Academic Press, 149–156.

Willmore, T. J. and Jhaveri, C., An Extension of a Result of Bang-Yen Chen, *Quart. J. Math. Oxford Ser.* 2, **23,** 319–323.

Wilson, J. P. (1966), The total absolute curvature of immersed manifolds, *J. Lond. Math. Soc.*, **40,** 362–366.

Wilson, J. P. (1969), Some minimal imbeddings of homogeneous spaces, *J. Lond. Math. Soc.* (2), **1,** 335–340.

Yang, P. and Yau, S. T. (1982) Eigenvalues of the Laplacian of compact Riemann surfaces and minimal submanifolds, to appear.

Yano, K. (1970), Integral formulas in Riemannian Geometry, Marcel Dekker, Inc.

Yau, S. T. (1982), Seminar on Differential Geometry, *Annal. of Maths. Studies*, **102,** Princeton.